Disaster Relief Aid

Bimal Kanti Paul

Disaster Relief Aid

Changes and Challenges

palgrave
macmillan

Bimal Kanti Paul
Kansas State University
Manhattan, KS, USA

ISBN 978-3-030-08418-9 ISBN 978-3-319-77282-0 (eBook)
https://doi.org/10.1007/978-3-319-77282-0

Cover credit: PJF Military Collection/Alamy Stock Photo

Printed on acid-free paper

This Palgrave Macmillan imprint is published by the registered company Springer International Publishing AG part of Springer Nature
The registered company address is: Gewerbestrasse 11, 6330 Cham, Switzerland

For those powerless in marginalized groups
often denied disaster assistance.

Foreword

Around 400 disaster events kill more than 100,000 people every year, with another 120 million people being directly affected. The international community annually spends $25 billion addressing these events which include floods, droughts, earthquakes, cyclones, wildfire, and similar. On a daily basis, people and communities have their lives taken or destroyed in catastrophic events that humans have little control over. Each night on our evening news, images are shared of buildings collapsed, farmlands cracked and parched, rivers swollen, or trees stripped of foliage by fierce winds. Throughout of all this, we see heroic efforts of people digging to free those trapped, feeding those that are starving, giving comfort to those who are bereft, or burying those that have been killed. Our experience of and response to these disasters is a common human experience and has been throughout our entire history.

Experiences of disasters have long been (and continues to be for many) understood in terms of "destiny" or "acts of God" that we have little control over. We simply endure these events as our lot in life. Indeed, it was not until the middle of the eighteenth century that philosophers, such as Voltaire and Rousseau, began to conceive disasters as the confluence between natural events and human activity.

On November 1, 1755, the great European city of Lisbon was devastated by a massive earthquake, tsunami, and fire. Estimates of those killed in the immediate aftermath of these events range between 10,000 and 100,000. However, the effects of this devastation were felt for many more years with long-term social, political, and economic ramifications resulting. In correspondence, Voltaire and Rousseau argued that such devastation could not be part of God's plan, nor could human activity (in this instance, overcrowded cities and poor planning) be ignored in understanding the effects of such natural events. In this manner, natural disasters came to be understood as outcomes arising from the nexus between the natural and human. As such, it was feasible that they could be planned for and in some regard, mitigated or have some vulnerability minimized.

By gaining this philosophical "control" over disasters, it was not unexpected that human responses in the aftermath of future natural events began to become increasingly organized as lessons were learned and resources to respond became more significant. At the same time, the slow but progressive process of globalization meant that news of disasters in one region or country become known more quickly in other parts of the world thus resulting in cross-border relief efforts increasingly occurring. Indeed, specific organizations were established with the primary purpose of providing support and resources to affected communities in both their own countries but also internationally. It was in this manner that the humanitarian sector was formed.

What drove these local and then, increasingly, international efforts was a desire to respond to human suffering. This humanitarian spirit (often linked explicitly to religious beliefs) underscored both the raising of funds and the enacting of responses. Over time, this humanitarian spirit became firmly entrenched alongside concepts of neutrality, impartiality, and independence in defining and guiding these humanitarian responses.

Throughout much of the history of humanitarian assistance, efforts to respond were led by volunteers, either those that dedicated their full-time efforts to such activities or those that were moved by individual events to offer their services in an ad hoc basis. While necessary and desirable, more recently though there has been a recognition that affected communities require more than good intentions in response. If the international community is to raise and expend tens of billions of dollars each year, then a level of professionalization is required to deliver effective, efficient, and appropriate responses that are locally appropriate and ideally, locally led. During the 1990s, standards were developed and adopted by the humanitarian sector in order to ensure more professional and accountable responses. The "Sphere Project" sets out the minimum standards which aid agencies responding to disasters should meet in planning and implementing their responses across a wide range of sectors, including health, education, sanitation, nutrition, and so forth. These standards were developed through aid agencies reflecting on their own practices and analysis of their own monitoring and evaluation.

Improving responses to disasters though also have resulted from careful research and analysis by academics interested in better understanding the nature of these disasters and how they are addressed, which brings us to this highly valuable book by Professor Bimal Paul. It is not unreasonable (nor unexpected) that humanitarian practitioners are focused on the immediate challenge of providing humanitarian assistance in circumstances of great pressure and stress. Their work, without exaggeration, is often a matter of life and death. Decisions they make immediately affect those who are the most vulnerable. Within this maelstrom of high demands, limited resources and tight time constraints, reflection and consideration of long-term trends and assessment of activities are a relegated priority. Yet, such consideration is vital in ensuring humanitarian assistance evolves and results in the most optimal outcomes as possible. It is therefore the work of researchers, such as Professor Paul, that is core to the ongoing professionalization of the humanitarian sector.

By utilizing an historical approach, Professor Paul is able to clearly map out the evolution of humanitarian responses and discuss the rapid "democratization" of disasters through the advent of social media. Such sharing of stories and needs has shortened the time between event and response, but also provided communities with new ways to seek assistance as well as provide immediate criticism where such assistance is less than satisfactory. Social media has also shaped the financial response of donors, both private and public. Understanding aid flows to disasters is very necessary for those seeking to maximize the resources they are able to access to deliver relief that is more comprehensive in terms of interventions funded or individuals reached.

Drawing on a range of case studies, Professor Paul provides clear insights and lessons on what are the main factors that result in disaster response being more or less effective. Again, in the immediate face of disasters, such analysis is crucial to improving humanitarian assistance. Indeed, it is the final chapter in this volume that is the greatest contribution. A range of challenges are recommendations that are made with the primary purpose of improving future responses.

Professor Paul's book is a very positive contribution to this field of humanitarian assistance. While there is increasing academic research in this field, it still remains under-considered in comparison with longer-term community development activities. More analysis is required and this volume signals a valuable addition. Drawing on a career-long interest in disaster and relief aid, this carefully constructed, well researched, and clearly written book pulls together a substantial amount of knowledge and expertise. This book will be read carefully by donors, aid agencies, and local community-based organizations.

Over the past nearly four years, my own academic institution has provided academic and professional training to nearly 400 humanitarian managers and leaders from across the globe. These dedicated and highly experienced professionals have responded to some of the world's most devastating disasters over the last three decades. Their actions have saved lives and allowed affected communities rebuild in the face of enormous hardship. During the course of their study, we ask them to reflect on their experiences and draw on this knowledge in their future responses. This volume provides a template for such analysis and will resonate with their own professional journey.

This volume is a valuable contribution to a field of study mired in the humanitarian spirit. This volume is well overdue and adds a considerable degree to our understanding of disasters and relief aid. As someone whose own research occurs in countries that are vulnerable to disaster events, I am personally grateful that this volume as the lessons it contains will enhance the humanitarian assistance provided by aid agencies.

Geelong, Australia
February 2018

Professor Matthew Clarke
The Centre for Humanitarian Leadership
Deakin University

Preface

In hazards and disasters research, I initially focused on issues and problems associated with disbursing humanitarian aid to disaster survivors or victims. Born and brought up in rural Bangladesh, I had frequently seen people anxiously waiting for emergency relief after natural disasters, which occurred more frequently in my home country during the 1960s and 1970s than now. I vividly remember disaster-affected, destitute people waiting for hours in long lines for food and other necessary items. Some of them happily carried relief goods to homes, but many others returned home either empty handed or with meager relief goods, complaining of biased aid distribution by relief agencies and local elites, politicians, and leaders. I debated then and do still whether disaster relief is the best answer to alleviate suffering of survivors caused by major and devastating environmental disasters. Later, when I conducted several field surveys on post-disaster public and private responses to extreme natural events in Bangladesh, Nepal, and the USA, I came to the conclusion that irrespective of country of origin, disaster survivors need, expect, and/or receive necessary emergency relief goods from external sources to mitigate the impacts of extreme events. Moreover, survivors often become proactive within their communities in terms of building networks and providing assistance to other survivors.

Disaster relief demands immediate action to ease the suffering of survivors and save lives. Although disaster relief is not designed to stop future disasters or reduce the risk of such events, it does help smooth the transition from necessary and immediate relief to both short- and long-term rehabilitation and redevelopment. A smooth transition helps survivors to return quickly to pre-disaster life so that they do not have to depend on any type of relief aid for an extended period of time. Moreover, rapid mobilization of relief supplies not only helps meet the immediate needs of the survivors, but also timeliness and effectiveness of disaster relief seem to be associated with the mental and physical health of survivors. Hazard and disaster researchers conclude that short-term relief operations may have longer-term implications for the material and emotional well-being of the survivors.

However, during my career, I have investigated many other important aspects of disaster: vulnerability and resilience, including how the public prepares for an impending environmental hazard; how victims receive and communicate warnings, as well as how they make decisions about protecting themselves against the threat of a destructive event; how survivors and public and private organizations respond to disasters; how survivors cope with and adapt to mitigate the damage and losses incurred during such events; and how they rebuild, recover, and return to the pre-disaster state. As a medical/health geographer, I also studied various health issues related to natural disasters, particularly circumstances, causes, and determinants of deaths and injuries caused by hurricanes/cyclones, earthquakes, tornadoes, and floods. I have studied these aspects of environmental hazards and disasters using rigorous scholarship and taken both theoretical and empirical approaches, which yielded useful practical and applied insights, including ways to increase the safety of people from the direct as well as indirect impacts of natural hazards. Consequently, my work associated with natural disasters has policy implications for increasing resiliency and reducing risk and suffering from such extreme events at individual, household, community, and national levels.

When I was approached by Palgrave in 2016 to write a book on disasters, I took time to respond primarily because I was preoccupied with completing our co-authored book, *Climatic Hazards in Coastal Bangladesh: Non-Structural and Structural Solutions* (published by Elsevier in 2017). Nonetheless, two issues inspired me to write the present book: my childhood memories of destitute people seeking disaster relief, which has natured my continued interest in relief operations. In the 1960s and 1970s, I also participated in relief operations in Bangladesh to distribute warm food among victims of tornadoes and floods. The second impetus came from my own publications on the topic in Bangladesh, India, Indonesia, Nepal, Pakistan, Sri Lanka, Thailand, and the USA. My sustained interest in disaster relief was facilitated by two grants from the National Science Foundation (NSF), one in 2015 (Rapid Response Research grant on the 2015 Nepal Earthquake) and the other in 2005 (Small Grants for Exploratory Research on Hurricane Katrina), and 10 Quick Response Grants from the Hazards Center, University of Colorado, Boulder. I have drawn heavily from these projects, particularly from the 2015 NSF grant. Thus, more disaster relief-related examples are cited from the 2015 Nepal Earthquakes.

The successful completion of this book was also contingent on the unconditional support I received from my colleagues and students in the Department of Geography at Kansas State University. I am particularly thankful to my geography colleague, Dr. Jeffery Smith, for his valuable comments on earlier drafts of each chapter of this book as well as his editorial suggestions. I also received editorial help from Katherine Krehbel, Marcella Reekie, and Nora Ransom, and I am thankful to them. My Ph.D. advisee, Avantika Ramekar, drafted all illustrations used in this book, and so I gratefully acknowledge her assistance. I would also like to thank Rachael Ballard, Head of Science and Society, Palgrave Macmillan, for her guidance and patience during the writing of this book. Finally, my deepest gratitude goes to my wife, Anjali Paul, daughters Anjana and Archana, and son Rahul, for their love, encouragement, and support over the years and particularly during the time I was busy writing this book.

Manhattan, USA Bimal Kanti Paul

Contents

Abbreviations

ACTED	Agency for Technical Cooperation and Development
ADB	Asian Development Bank
ADRA	Adventist Development and Relief Agency
AfDB	African Development Bank
AIDS	Acquired Immuno Deficiency Syndrome
ALNAP	Active Learning for Accountability and Performance
AP	Associated Press
ARA	American Relief Administration
ARC	American Red Cross
CAP	Consolidated Appeal Process
CARE	Cooperative for Assistance and Relief Everywhere
CDEM	Civil Defense Emergency Management
CERF	Central Emergency Response Fund
CNN	Cable News Network
COVOAD	Colorado Voluntary Organizations Active in Disasters
CRB	Commission for the Relief of Belgium
CRED	Center for Research on the Epidemiology of Disasters
CRS	Catholic Relief Services
DAC	Development Assistance Committee
DaLA	Damage and Loss Assessment
DCs	Distribution Centers

DEC	Disasters Emergency Committee
DHS	Department of Homeland Security
DHSEM	Department of Homeland Security and Emergency Management
DKKV	German Committee for Disaster Reduction
DNF	Dalit NGO Federation
DOD	Department of Defense
EC	European Community
ECLAC	Economic Commission for Latin America and the Caribbean
EEA	European Environment Agency
EF-scale	Enhanced Fujita-scale
EM-DAT	Emergency Events Database
EOP	Emergency Operation Plan
ERC	Emergency Relief Coordinator
EU	European Union
FAO	Food and Agriculture Organization
FEMA	Federal Emergency Management Agency
FGMs	Focus Group Meetings
FTS	Financial Tracking System
GAM	Gerakan Aceh Merdeka (Free Aceh Movement)
GDP	Gross Domestic Product
GFDRR	Global Facility for Disaster Reduction and Recovery
GHD	Good Humanitarian Donorship
GHP	Global Humanitarian Platform
GoB	Government of Bangladesh
GoN	Government of Nepal
GWWC	Giving What We Can
HHI	Harvard Humanitarian Initiative
HL	Human Logistics
HP	High Priority
HRC	Hazards Research Center
HRNA	Human Recovery Needs Assessment
HRPs	Humanitarian Response Plans
IASC	Inter-Agency Standing Committee
ICRC	International Committee of the Red Cross
IDB	Inter-American Development Bank
IDNDR	International Decade for Natural Disaster Reduction
IDPs	Internally Displaced Persons

IDSN	International Dalit Solidarity Network
IFRC	International Federation of Red Cross and Red Crescent Societies
IGAD	Intergovernmental Authority on Development
IGOs	Intergovernmental Organizations
IMF	International Monetary Fund
INGOs	International Nongovernment Organizations
IOT	Indian Ocean Tsunami
IRIN	Integrated Regional Information Network
ISCU	International Save the Children Union
ISDR	International Strategy for Disaster Reduction
JDNLA	Joint Damage, Loss, and Needs Assessment
JeI	Jamaat-e-Islami
JuD	Jamaat-ud-Dawa
KPK	Khyber Pakhtunkhwa
LDCs	Local Distribution Centers
LeT	Lashkar-e-Taiba
LoC	Line of Control
LP	Low Priority
LRCS	League of Red Cross Societies
LSPs	Logistic Service Providers
MCC	Mennonite Central Committee
MINUSTAH	Mission de Nations Unies pour la Stabilisation en Haïti
MMI	Majelis Mujahideen Indonesia
MoFDM	Ministry of Food and Disaster Management
NAO	National Audit Office
NATO	North Atlantic Treaty Organization
NCDHR	National Campaign on Dalit Human Rights
NGOs	Nongovernmental Organizations
NP	Nonpriority
NPR	Nepalese Rupees
NSF	National Science Foundation
NSTC	National Science and Technology Council
NWFP	North West Frontier Province
OCHA	Office for the Coordination of Humanitarian Affairs
ODA	Official Development Assistance
ODI	Overseas Development Institute
OECD	Organization for Economic Cooperation and Development

OECD/DAC	Development Assistance Committee of the Organization for Economic Cooperation and Development
OFDA	Office of Foreign Disaster Assistance
OHA	Official Humanitarian Assistance
OPEC	Organization of Petroleum Exporting Countries
OSM	OpenStreetMap
OXFAM	Oxford Committee for Famine Relief
PAHO	Pan American Health Organization
PD-HL	Post-Disaster Human Logistics
PDNA	Post-Disaster Needs Assessment
PDNA/RF	Post-Disaster Needs Assessment and Recovery Framework
PPP	Public–Private Partnerships
PTSD	Post-traumatic Stress Disorder
RC/HC	Resident Coordinator or Humanitarian Coordinator
RFIDs	Radio Frequency Identification Devises
R-HL	Regular Human Logistic
RHRP	Rural Housing Reconstruction Plan
RNA	Royal Nepalese Army
RRPs	Refugee Response Plans
SAPs	Structural Adjustment Programs
SAR	Search and Rescue
SCF	Save the Children Fund
SDI	Susila Dharma International
SIDS	Small Island Developing States
SPLM	Sudan People's Liberation Movement
SV	Spontaneous Volunteers
TIB	Transparency International Bangladesh
TLA	Tamil Liberation Army
U.N.-ESC	United Nations—Economic and Social Council
UNCHS	United Nations Center for Human Settlement
UNDP	United Nations Development Program
UNDRC	United Nations Disaster Relief Coordinator
UNICEF	United Nations Children's Fund
UNIDS	Joint United Nations Program on HIV/AIDS
UNMISS	United Nations Mission in South Sudan
USAID	United States Agency for International Development
USAR	Urban Search and Rescue

V&TCs	Volunteer and Technical Communities
WASH	Water, Sanitation, and Hygiene
WFP	World Food Program
WHO	World Health Organization

List of Figures

List of Tables

1

Introduction

Natural disasters pose a serious threat to the well-being of affected individuals, households, communities, and countries. Those who suffer and survive such events, particularly major events, cannot cope without emergency assistance from external sources, both within and beyond the national boundary. Similarly, the local governments of affected communities depend on these sources because they cannot provide food, medicine, and other basic necessities for those who survived the disasters without generous help from such sources. Outside sources include people, governments, and a multiplicity of organizations not only from unaffected areas of the country, but also from many entities from foreign countries. At the country level, the government of the affected country frequently relies on foreign sources for emergency assistance to tackle challenges in the aftermath of a major natural disaster. This is particularly true for the developing countries. Thus, the availability of foreign sources of funding, including in-kind, cash, and other help, is crucial to developing countries because governments of these countries are least likely to have the capacity and resources to successfully organize disaster relief.

Disaster relief is defined by the World Health Organization (WHO) as "The provision of assistance during or immediately after a disaster to

B. K. Paul, *Disaster Relief Aid*,
https://doi.org/10.1007/978-3-319-77282-0_1

meet the life preservation and basic subsistence needs of those people affected. It can be of an immediate, short-term, or protracted duration" (WHO 2007, 31). According to New World Encyclopedia (2017), "Disaster relief is the monies or services made available to individuals and communities that have experienced losses due to disasters such as floods, hurricanes, earthquakes, drought, tornadoes, and riots." Both are simple definitions that do not encompass the broad spectrum of activities, processes, and mechanisms, and many actors, agencies, and stakeholders involved in collecting, sorting, transporting, storing, and disbursing emergency assistance to survivors of extreme events.

However, because so many humanitarian actors and other stakeholders become involved, disaster relief operations as a whole are quite complex. Without clear mandates, roles, objectives, coordination, and accurate information, such operations are at risk. According to Barbarosoglu et al. (2002, 118), the core focus of disaster relief operations is to "design the transportation of first aid material, food, equipment, and rescue personnel from supply points to a large number of destination nodes geographically scattered over the disaster region and the evacuation and transfer of people affected by the disaster to the health care centers safely and very rapidly."

In this book, the terms emergency relief, emergency assistance, emergency aid, disaster relief, disaster relief operation, disaster aid, humanitarian assistance, and humanitarian aid are used interchangeably. Although some authors (e.g., Margesson 2005) include search and rescue, and treatment of injured survivals as parts of relief operations, here emphasis will be on the collection, transportation, store, and delivery of relief goods to the victims of natural and man-made disasters.

Disaster assistance should provide rapid relief from humanitarian hardship after a natural disaster. Obviously, the more severe the disaster, the more aid must flow to counter its effects. Humanitarian aid includes all actions designed to save lives, alleviate suffering, and maintain and protect human dignity during and after both man-made crises and natural disasters, as well as to prevent and strengthen preparedness for their occurrence. The United Nations Office for the Coordination of Humanitarian Affairs (OCHA) and Vodafone Foundation (2011) define humanitarian aid as "[a]n intervention to help people who are victims of a natural disaster or conflict meet their basic needs and rights." According to the guiding principles stated in the U.N. resolution A/RES/46/182, "[h]umanitarian assistance must be provided in accordance with the principles of humanity, neutrality and impartiality" (U.N. 1991). This means that different actors taking part in a disaster relief operation must comply with these three principles. Like the U.N. resolution, the International Red Cross and Red Crescent Movement states seven fundamental principles: humanity, impartiality, neutrality, independence, voluntary service, unity, and universality (IFRC 2013).

In essence, all humanitarian actions should be guided by the U.N. stated humanitarian principles of humanity, impartiality, and neutrality. *Humanity* means the centrality of saving human lives and alleviating suffering wherever it is found. *Impartiality* means implementation of actions based solely on need, without any form of discrimination (as to nationality, race, religion, beliefs, class, or political opinions) between or within affected populations. In the context of *neutrality*, if there is any conflict or dispute, the humanitarian organization should not support one group over the other especially where aid is offered. The International Red Cross and Red Crescent Movement's principle of—*independence* means humanitarian objectives are autonomous, free of political, economic, military, or other objectives of any actor in areas where humanitarian action is being implemented (Good Humanitarian Donorship 2003; IFRC 2013).[1] *Voluntary service* means relief efforts must not prompted by desire for gain. The remaining two principles (unity and universality) are applicable to the organization. Under *unity*, the organization must carry its relief operation throughout its territory, and *universality* indicates that all Red Cross Societies "have equal status and share equal responsibilities and duties in helping each other ..." (IFRC 2013, 262).

It is worth mentioning at the outset that the focus of this book is on humanitarian relief primarily intended for sufferers of natural disasters, not victims of "complex" or man-made emergencies. Complex emergencies are those humanitarian crises that occur in a country, region, or society where internal or external conflict has caused total or considerable breakdown of authority. These crises require an international response that goes beyond the mandate or capacity of any single agency and/or the ongoing U.N. country program (IASC 1994).[2] Such "complex emergencies" are typically characterized by (1) extensive violence and loss of life; (2) displacement of populations; (3) widespread damage to societies and economies; (4) the need for large-scale, multifaceted humanitarian assistance; (5) hindrance or prevention of humanitarian assistance because of political and military constraints; and (6) significant security risks for humanitarian relief workers in some areas (IASC 1994).

Background on Humanitarian Assistance

Relief given immediately after a natural disaster or complex emergency to persons and households who have suffered, incurring damage to material goods and property from an extreme event, has a long tradition (Kent 1987). To many, post-disaster relief efforts, along with generous support to victims of human-created emergencies or complex humanitarian emergencies, dates back to the pre-colonial period when local rulers and wealthy persons provided short-term support after natural disasters and/or complex emergencies. This humanitarian relief has largely been driven by the sense of moral obligation among people not affected by the disasters to respond and address the suffering (Bradshaw 2013).[3] During the colonial period, colonial rulers and Christian missionaries led an expansion of relief efforts beyond the local level to include all sufferers of disasters or conflicts, regardless of country of origin, race, religion, caste, or culture. The principal aim for such a gesture was to help humanity (Bradshaw 2013).

During the colonial period as well as after the two World Wars, several notable humanitarian organizations were founded and contributed to spreading disaster relief across the globe. The International Committee of the Red Cross (ICRC) was formed in Geneva, Switzerland, in 1863 in the aftermath of the Battle of Solferino in northern Italy in 1859. The committee was founded by Swiss businessman Henry Dunant (Kent 1987; Alexander 2011). The sole objective of the ICRC was to ensure effective medical aid and assistance to injured soldiers and other survivors and sufferers of civil unrest and armed conflict. In 1914, the Commission for the Relief of Belgium (CRB) was formed to address the food needs of Belgian and French civilians in territories occupied by Germany that were subject to blockade by Allied forces (Davey et al. 2013). These institutions inspired the formation of the League of Red Cross Societies (LRCS) in 1919 in Geneva, Switzerland, the forerunner to the International Federation of the Red Cross and Red Crescent Societies (IFRC). In 1983, the LRCS was renamed the League of Red Cross and Red Crescent Societies and finally the International Federation of Red Cross and Red Crescent Societies in 1991 (Davey et al. 2013).

The devastation of the First World War also prompted the birth of the first transnational humanitarian non-governmental organization (NGO), the Save the Children Fund (SCF) or simply Save the Children (Walker and Maxwell 2009). This organization was formed in Britain in 1919 with a primary mission to improve the lives of children by providing health care, economic opportunities, and better education. Also, all children, including the children of former enemies, were eligible for emergency aid in natural disasters, wars, and other conflicts (Freeman 1965). As more national SCF sections were established in different European countries, its leaders, Eglantyne Jebb and Dorothy Buxton, oversaw the formation of the International Save the Children Union (ISCU) in Geneva in 1920, with the British SCF and the Swedish Rädda Barnen as its leading members (Alexander 2011). The American Relief Administration (ARA) was formed in 1919. Unlike SCF, however, ARA was subject to US government control. In 1942, the Oxford Committee for Famine Relief (Oxfam) was formed in the UK, and Cooperative for Assistance and Relief Everywhere (CARE) was founded in the USA in 1945 (Davey and Scriven 2015). Other major relief organizations are listed in Table 1.1 by name, year of establishment, and country of origin.

An extensive humanitarian relief community has developed since the Second World War (Therien and Lloyd 2000; Oloruntoba and Gray 2006). Randolph Kent (1987) claims that World War II was as a turning point for greater international intervention on behalf of disaster-stricken peoples.[4] Similarly, Barnett and Walker (2015) also maintain that relief agencies, which they termed the "Humanitarian Club,"

Table 1.1 Selected major relief organizations by name, year of establishment, and country of origin

Name	Year	Country of origin
Salvation Army	1865	Britain
Pan American Health Organization (PAHO)	1902	USA
Mennonite Central Committee (MCC)	1902	USA
Catholic Relief Services (CRS)	1943	USA
Lutheran World Federation	1947	Sweden
Christian Aid	1949	Britain
World Vision International	1950	USA
Disasters Emergency Committee (DEC)	1963	Britain
Concern Worldwide	1968	Ireland
Doctors Without Borders (Medecins Sans Frontieres)	1971	Switzerland
Action Aid	1972	Britain
Muslim Aid	1985	Britain

Source Prepared by the author

born after World War II as a Western response to the challenges of reconstructing Europe and decolonizing its former territories. The war also marked the shift in the focus of relief from developed to developing countries.

Humanitarian organizations proliferated after the Second World War. In addition to organizations established during the war, nearly 200 international NGOs (INGOs) were formed between 1945 and 1949, most of them in the USA (Barnett 2011). Increasing numbers of humanitarian organizations were spurred by the severe famines that occurred in Europe (e.g., the Poland Famine, 1940–1945; the Dutch Famine, 1944–1945; and the Greece Famine, 1941–1944); Africa (e.g., Morocco, 1940–1948; and the Rwandan Famine of 1944); and Asia (e.g., the Bengal Famine, 1943–1944; the Chinese Famine of 1942–1943; and the Indochina Famine of 1945) during the first half of the twentieth century (Davey et al. 2013). Among the most notable of these are the Bengal Famine of 1943 and Indochina Famine of 1945. The first was more severe, causing the deaths of approximately 3 million people to starvation and related diseases like malaria, cholera, and smallpox (Greenough 1980; Devereux 2000). This and other famines led the formation of the Food and Agriculture Organization (FAO) of the United Nations in 1945 with the aim of defeating world hunger and later the

World Food Program (WFP) in 1961 (Barrett and Maxwell 2005; Alexander 2011).

With decolonization, which began after World War II and had nearly ended by the late 1960s, the governments of newly independent countries faced severe economic hardships as well as a lack of resources, experience, and infrastructure. For these reasons, they could not provide emergency assistance to survivors of major natural disasters without the financial assistance of external sources, including governments of developed countries (UNDP 2004). Thus, before the 1970s, in addition to development aid, the governments of disaster-affected developing countries also received generous disaster assistance from foreign governments, and other humanitarian actors like international agencies and INGOs, as well as private domestic organizations, both charitable and faith-based, and local and national volunteer groups (UNDP 2004). While foreign governments and international humanitarian organizations generally channeled disaster aid through governments of disaster-affected countries, the local and national charitable institutions and volunteers directly distributed disaster relief to survivors. Foreign governments also often delivered emergency assistance via the United Nations (IFRC 1997).

Except for the government of any disaster-affected developing country, all other emergency assistance providers are here termed humanitarian actors or agencies. There are two types of actors: international or foreign and domestic or local actors. As indicated, before the 1970s, most emergency aid flowed either directly from humanitarian actors to disaster survivors or indirectly through the government of the affected country. International actors often sent some part of the aid to the U.N., which, in turn, delivered that aid to the government of the affected country.

However, since the early 1970s, development partners increasingly realized that natural disasters are a primary obstacle to the development of developing countries. Thus, they started to link disaster relief, rehabilitation, prevention, and preparedness to development and channeled both types of aid exclusively to the governments of developing countries. The basic premise was to create synergies between short-term relief measures with long-term development programs.[5]

Unfortunately, since the early 1980s, many governments of developing countries show consistent failure in properly managing and delivering relief goods to disaster survivors. Thus, emergency efforts and rehabilitation programs were ineffective because of widespread irregularities and corruption, as well as public and private interference (Paul 1998). After noticing this trend, INGOs, foreign faith-based organizations, and foreign corporate businesses, governments, and volunteers (i.e., external emergency aid sources) started to disburse aid either directly to the sufferers of natural disasters or indirectly through national NGOs and local agencies or local actors (Paul 1998, 2006, 2011).

The inefficiency in the governments of many developing countries provided an opportunity for national, regional, and local NGOs to grow rapidly in these countries during the 1980s. This growth was also the result of the inefficiency of a top-down, command-and-control, or technocratic approach to disaster relief distribution. In addition, an influential global consensus arose, spurred by the international financial institutions of the World Bank and International Monetary Fund (IMF), for the need for political democracy and good governance on the one hand and economic liberalization on the other. Policies of political and economic liberalization typified by the Structural Adjustment Programs (SAPs) of the IMF have become the new development doctrine for the Global South. Within this discourse, NGOs, as a group, are heralded as a development panacea (Booth 1994; Watts 1993). Table 1.2 provides a general list of important strengths and weaknesses of NGOs.

The United Nations International Decade for Natural Disaster Reduction (IDNDR), which started in 1990 and ended in 1999, also urged that NGOs become involved in disaster reduction efforts. The IDNDR appealed for mobilization of NGOs and the participation of local communities in disaster relief operations (Wisner et al. 2004). In fact, considerable effort involved these two entities during the second half of the IDNDR's existence. However, the shift in channeling disaster aid shows up in funding statistics: in 1976, no European Community (EC) emergency aid funding went through NGOs; by 1982–1983, NGOs received 40% of all aid funding (Borton 1993). By that time, national NGOs showed an understanding of local conditions

Table 1.2 Strength and weaknesses of NGOs

Strengths
1. NGOs are less bureaucratic, swift in decision making, and has low overhead expenses
2. NGOs generally work on a small-scale grassroots project in remote and isolated areas, be it in development or disaster relief
3. NGOs' rapport with villagers allows them to draw on local knowledge systems in the design of technology options and to strengthen them
4. NGOs can design location specific, innovative, disseminate method on specific commodity/practice
5. Group formation by NGOs is effective, and strong supervision can be maintained. NGO activities and systems are easy to replicate
6. NGOs are flexible, adaptive, and participatory
7. Innovative work can be pushed comfortably
8. NGOs' outreach capacity is more effective to work with grassroots people. They are good at identifying vulnerable groups
Weaknesses
1. Activities of some NGOs are more sporadic, independent, small in size, more action-oriented and short-ranged rather than long-ranged and learning-oriented
2. Most NGOs lack the resources and are dependent to external funding sources
3. NGOs are not accountable to others including the government
4. NGOs are not obligated to do partnership or collaborative work with government and even with other NGOs
5. Absence of self-evaluation mechanism
6. High job insecurity and high turnout

Source Compiled by the author

far higher than the governments of disaster-stricken developing countries as well as delivering relief aid faster than the government bureaucrats. In 2005, between 48 and 58% of all known humanitarian funding flowed through NGOs (Walker and Pepper 2007).

Development aid as well as disaster aid is also now channeled through NGOs of developing countries, particularly ones characterized by weak institutions and policies (Raschky and Schwindt 2010). For example, in 1972–1973, less than 1% of total foreign aid was channeled through NGOs in Bangladesh. This increased to 16% by 1985–1986 and 22% by the end of the 1990s (Alam 1988; White 1991; Streeten 1999).

Because of this trend, national governments of developing countries have had less and less success in attracting both emergency assistance and developmental aid from external sources (Townsend et al. 2004). This change was necessary because domestic NGOs have proven over the years that they are more efficient, effective, and impartial in implementing of development programs as well as in disbursing of relief than most government agencies (Fowler 1990; Lewis 1992; Paul 2006). Because domestic NGOs do participate in disaster relief efforts, relief distribution among disaster survivors in many developing countries has been almost flawless (Paul 2011). Government entities in developing countries, however, tend to downplay the role of the NGOs in relief efforts.

At the same time, foreign donors and development agencies have shifted their focus more toward relief and away from development assistance: WFP, for instance, cut its development projects from half its activity in 1989–1990 to less than one-sixth by 2000 (Clay 2003). This led national NGOs to assume the responsibility for delivering disaster relief assistance to survivors to extreme natural events, a role traditionally assumed by the national government of developing countries affected by disasters (Paul 2006). The massive growth of information technology and mass communication, along with the growth of world relief, in large measure represented by U.N. efforts, the Red Cross and Red Crescent Societies, and NGOs in the 1980s and early 1990s, has transformed what were once essentially regional and local problems onto the world stage (Borton 1993; Alexander 2006). Thus, the globalization of disaster relief corresponds with the emergence of modern economic globalization, which is driven by international trade and foreign direct investment and aided by information technology and the reduction in trade barriers.

Figure 1.1 shows the general flow of disaster aid from international humanitarian sources to disaster survivors of developing countries between 1950 and 2000. Between 1950 and the early 1970s, the flow came from external sources to disaster survivors via the governments of disaster-affected countries. In the second phase (the mid-1970s to early/mid-1980s), NGOs appeared for the first time as distributors, becoming the principal distributors in the third phase (mid-1980 to 2000). In recent years, NGOs have started to behave like government agencies of

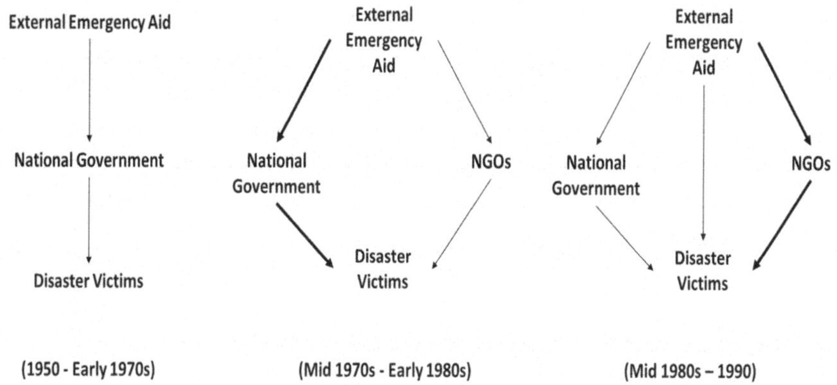

Fig. 1.1 Schematic of relief flow from external emergency sources to disaster victims from 1950 to 1990

the 1970s and 1980s and thus have lost their role as the dominant distributors of disaster relief. These organizations are no longer considered efficient, effective, and impartial in disbursing disaster relief (Islam and Walkerden 2015).[6] For these reasons, NGOs lost their previous position as dominant distributors of development aid and relief aid (Trivunovic et al. 2011).

In the late 1990s and early 2000s, the governments of many developing countries realized that domestic NGOs were replacing them in their traditional role of providers and distributors of disaster aid. They became desperate to regain their authority and control over external disaster assistance. To reach this goal, many governments of developing countries have started distributing emergency aid more fairly and impartially and have become more accountable and transparent (Paul 1998, 2003, 2006). Unlike the past, the mass media and concerned citizens of developing countries are more vocal in criticizing national governments for improper emergency relief distribution. This criticism helped governments to provide disaster relief as efficiently as possible. The goal seemed to be to impress external sources by effectively distributing the emergency relief aid.

It is worthwhile to mention that by now, a host of international emergency aid agencies or INGOs like Action Aid, CARE International,

Oxfam, Save the Children, and World Vision have been working in many developing countries for decades. These agencies either have local staff, or partner with local aid organizations in disaster-affected developing countries and often directly distribute overseas humanitarian assistance. Some of these agencies also often collaborate with governments of countries affected by disaster to provide relief in the immediate aftermath of disasters.[7]

Humanitarianism and Volunteerism

Closely associated with humanitarian assistance or relief are the terms humanitarianism and volunteerism. These three terms have more similarities than differences, and their origins date to about the same time; they also all serve the same purpose. Humanitarianism is an active belief in the value of human life whereby humans assist others who need assistance. In essence, humanitarians seek to promote human welfare. Humanitarianism is based on the principle that, irrespective of belief, race, religion, ancestry, or nationality, all human beings deserve respect and dignity. Therefore, humanitarians work voluntarily to promote the well-being of humanity as a whole in a wide range of activities (Chimni 2000). In the context of disasters, whether natural or man-made, they save lives and alleviate the sufferings of victims. Humanitarianism dictates that every victim of the event has the universal right to receive humanitarian relief.

Humanitarians either work alone or formally as volunteer members of humanitarian (relief) organizations. Humanitarianism is generally an individual philosophy, but the ideology of humanitarianism promotes unification. It "embraces individuals in a collective identity, irrespective of the differences and divisions that may separate them" (Thompson 1990, 64). It is global in scope and draws the world together (UNDP 1999).

Although humanitarianism does not emphasize any religion, it originated in the West, deriving from Christian missionaries who were instrumental in bringing kindness and goodwill to the developing world.[8] Humanitarianism is an apolitical philosophy, but it has never been divorced from (international) politics. Humanitarian aid has always been used international relief agencies to manipulate relief for political ends (Chimni 2000; Duffield et al. 2001; Fox 2001). With the end of the Cold War, the rise of global civil society and modern globalization, and a growing North–South divide in wealth and technology, a new humanitarianism has developed. It differs from the earlier versions of humanitarianism by rejecting the traditional principles that guided humanitarianism through the last century, promoting instead human rights and using aid to secure peace. "New humanitarians reject the political naivety of the past, assess the long-term political impact of relief and are prepared to see humanitarian aid used as a tool to achieve humanrights and political goals" (Fox 2001, 275). The universal right to unconditional humanitarian relief in times of crisis seems a good place to start in defining the new set of humanitarian principles. However, practice does not always follow principle, and the future will judge whether the past and new humanitarianism actually differ.

In the case of natural disasters, local people, who have no formal training in emergency management, are usually the first responders during such extreme events. They participate in search and rescue (SAR) operations before the domestic and international professional SAR teams arrive and provide emergency assistance to disaster survivors and sufferers before formal relief operations begin. In addition to the many organizations participating in such relief operations, another group of people come to disaster sites as individuals, either formally through nonprofit and government agencies or informally through local communities and online networks. These are volunteers from non-affected areas and other countries, people dedicated to helping disaster survivors and reducing their sufferings. These volunteers, often called spontaneous volunteers (SV), may or may not have appropriate training or experience, but they volunteer their time freely to benefit another person, group, or organization (Barraket et al. 2013). In their integrated theory of volunteer work, Wilson and Musick (1997) maintain that volunteer work is productive work that draws on human capital. They further maintain that volunteer work reflects collective behavior that uses social capital and ethically guided work that draws on cultural capital.

In volunteerism research, the term generally refers to non-obligatory or unpaid activities undertaken in an organized way for the benefit of others through groups or philanthropic organizations (Whittaker et al. 2015). Volunteerism has numerous and differing definitions as do activities constituting volunteerism and who can be considered a volunteer. Cnaan et al. (1996) reviewed 300 definitions of volunteerism and posited that these definitions generally comprise four common dimensions: the extent to which an act is voluntary (ranging from free will to obligation), the nature of the reward or remuneration (if any), the structure within which volunteer activities take place (informal to formal), and the intended beneficiaries of volunteering (ranging from strangers to the volunteer themselves) (also see Barraket et al. 2013). According to the strictest definitions, volunteerism must "be entirely voluntary and entail no coercion; involve no reward or even personal interest in the voluntary activity; be undertaken through a formal organization; and involve no relationship or similarities (e.g. ethnicity) between volunteers and beneficiaries" (Whittaker et al. 2015, 360). The strict definitions exclude volunteers who converge at disaster sites without any affiliation with any formal organization and are silent about an important dimension in volunteerism: time contribution.

Unlike such strict definitions, broader definitions of volunteerism encompass activities that "involve degrees of coercion (e.g. volunteering as part of a school programme); remuneration below the value of work and services provided; are undertaken outside of formal organisations; involve people of similar backgrounds (e.g. ethnic, religious, gender or residential groups) and even volunteers as beneficiaries (e.g. self-help groups)" (Whittaker et al. 2015, 360). It is to be noted that volunteerism is defined more broadly in emergency and natural disaster contexts. However, most definitions of volunteerism in the context of natural disasters do not include volunteers who are not affiliated with relief organizations For example, the US Federal Emergency Management Agency (FEMA) defines volunteer as "any individual accepted to perform services by lead agency (which has authority to accept volunteer services) when the individual performs services without promise, expectation, or receipt of compensation for services performed" (cited in Whittaker et al. 2015, 361).

These and most other definitions of volunteerism in the context of natural disasters include only formal volunteers, but the informal, unofficial, or unaffiliated volunteers have participated increasingly in recent years during disaster relief efforts. For this reason, the Civil Defense Emergency Management (CDEM) in New Zealand includes them in its definition. The CDEM defines spontaneous volunteers as "unaffiliated – not a member of an organized or recognized group, untrained in CDEM relevant skills, not responsible or accountable to any given organization, an unknown quantity, possibly a 'previous' volunteer (no longer involved)" (CDEM 2006). This definition includes emergent volunteers and convergent volunteers. Washington County Emergency Operation Plan (EOP), Oregon, USA (2011), defines an emergent (unaffiliated) volunteer as "an individual who either calls to offer assistance or arrives at a disaster scene and either is not affiliated with a recognized disaster agency, or chooses not to volunteer for the agency with which they are affiliated and who is assigned one or more volunteer tasks."

An affiliated volunteer, on the other hand, is an individual who has completed the prerequisites to be officially recognized by, and has registered with a recognized agency or organization. A person may be an affiliated volunteer with a particular agency and an emergent volunteer with another agency (Washington County Emergency Operation Plan 2011). A convergent volunteer is a person who arrives at the disaster site unexpectedly to render aid to survivors of extreme events. Once on the site, such personnel generally engage in freelancing and operate at an emergency incident without knowledge of or direction by the on-scene command authority (Cone et al. 2003).

In contrast to the definition of spontaneous volunteers provided by CDEM (2006), the Australian Red Cross (2010) definition does not include people who provide assistance at a disaster site under their own initiative. Rather it divides spontaneous volunteers into two categories: potential spontaneous volunteers and casual volunteers. The former are people who express an interest in volunteering after a disaster, while casual volunteers are those who have been inducted and screened, though they remain unaffiliated with the emergency service (Barraket et al. 2013). The potential spontaneous volunteers were not previously affiliated with any relief organization and may or may not have the required skills for emergency work (Drabek and McEntire 2003).

In a recent study, Whittaker and his colleagues (2015) categorized informal volunteerism into two broad types: emergent volunteerism and extending volunteerism. They also identified digital volunteerism as a new mode of volunteerism, driven by the increasing accessibility to information technologies. The digital volunteerism may be emergent or extending. Whittaker et al. (2015, 363) define extending volunteerism as "Groups and organisations without emergency or disaster functions [who] often extend their activities to volunteer in times of crisis." They work as a group and generally represent part of an existing community group such as a sporting club, Community Women Association, and religious or service organization. These volunteers often have clear knowledge and can draw on existing resources and networks to meet them.

Digital volunteers are a new breed of people willing to help disaster survivors with their technical skills, often without going to the disaster site. As a group, they use a computer, social media, or mobile devices to share information or digital maps that can be very helpful for first responders and others involved in rescue and relief operations. Because of the availability of technology, digital volunteerism is likely to become increasingly prevalent in post-disaster response and relief activities throughout the world. One strength of this volunteerism is that volunteers do not necessarily have to spend long hours participating in activities associated with emergencies (Whittaker et al. 2015).

Criticisms of Disaster Aid Provision

Hazard researchers have debated whether disasters should be addressed at all by relief measures. As responsibility shifted for organizing and distributing emergency assistance for survivors of disasters, a number of hazard researchers (e.g., Susman et al. 1983; Blaikie et al. 1994; Hewitt 1997; Bolin and Stanford 1999) voiced opposition, suggesting several reasons that survivors of disasters should not receive relief at all. These researchers maintain that disaster relief is counterproductive; it can increase dependence, creating a "culture of dependency," continuing the underdevelopment and marginalization of disaster survivors in developing countries (World Bank 2000). Relief dependency occurs when disaster survivors lose their initiative to fend for themselves and become reliant on external assistance for their near-future livelihood. Some authors (e.g., Harvey and Lind 2005; Islam and Walkerden 2015) suggest that relief dependency is a logical coping strategy for disaster survivors, given the uncertainties of future access to relief and the costs of the alternatives such as securing loans from moneylenders or microcredit organizations. Harvey and Lind (2005, 1) maintain that: "the focus should not be on avoiding dependency, but on providing sufficiently reliable and transparent assistance so that those who most need it understand what they are entitled to, and can rely on it as part of their own efforts to survive and recover from crisis."

However, critics of disaster relief further maintain that, like emergency aid, development aid and development programs are often to blame for increasing human exposure to environmental risk. They view disasters a consequence of development, urbanization, and industrialization. For examples, both global warming and climate change are widely considered the result of increased emissions of greenhouse gases in the atmosphere, which, in turn, is primarily caused by burning fossil fuels in industrialized and developed countries of the world (Rashid and Paul 2014).

Another example link between development and disaster was exposed in the devastating 2016 flood in the Baton Rouge area, Louisiana, USA. This flood killed 13 people, displaced tens of thousands more, caused an estimated $8.7 billion in damage, and destroyed some 60,000 houses. Experts maintain that poorly regulated suburban growth in Baton Rouge that are prone to flooding may have contributed to the 2016 flood. Flood water, the result of heavy rainfall in August, did not drain quickly because of man-made structures like new roads, houses, bridges, or levees.[9] These structures either obstructed or slowed the natural drainage pattern, creating a 100-year flood (Ball 2016). For the same reason, a catastrophic flooding event occurred in Houston, Texas, following the record-breaking amount of rainfall that accompanied by Hurricane Harvey in August 2017 and caused around $180 billion (Times-Picayune 2017).[10] However, Louisiana has not seen a natural disaster of this magnitude since 2005, when the storm surge from Hurricane Katrina submerged much of New Orleans and the surrounding areas. Similarly, construction of dams and levees along the Mississippi River basin encouraged denser settlement patterns and industrial development in flood-prone areas. This, in turn, inflicted much greater losses in a large flood that caused the levees to fail.

In providing disaster aid, Susman et al. (1983, 279) claim that "relief usually reinforces the *status quo ante*, namely the process of underdevelopment that produced such vulnerability in the first place." "Even in 'normal' times the poorest sections of society are pressured to over-use the land and, when disaster strikes, the conventional responses merely accelerate the continued underdevelopment and marginalization" (Smith 1992, 44). Critics like Susman et al. (1983), Sollis (1994), and Wisner et al. (2004) maintain that the goal of disaster relief and recovery efforts is returning disaster survivors to their pre-disaster, normal life. However, these researchers question what good it is to return to a pre-disaster life characterized by injustice, extreme poverty, and disempowerment. They do not believe disaster relief can be effective without social and political change (Mustafa 2003).

Critics of disaster relief recommend sustainable solutions to disaster problems instead of repeating the same relief operations. They urge that relief should incorporate the goals and mechanisms of sustainable development into the paradigm of relief. This will better alleviate the structural conditions of pre-disaster vulnerability that predispose people to experience disaster or that increase the severity of a disaster's impact (e.g., Cuny 1983; Kibreab 1987; Slim and Mitchell 1992; Zetter 2003).

Disaster relief is often reactive, counterproductive, crisis-oriented, and not usually long term and/or proactive. Relief efforts are externally managed and non-participatory. Such efforts fail to recognize the power of local institutions, knowledge, insight, or skills and adopt top-down strategies that preclude local participation (e.g., De Waal 1997; Platt 2000). These efforts had often been criticized for ignoring, sidelining, or actively undermining local efforts (Bragg 2015). For example, in 1972, 30 nights of ground frost occurred in the highlands of Papua New Guinea (Kent 1987). Outside relief officials residing in the country organized a massive relief effort anticipating a famine but completely ignoring how the local people could cope. "All the available evidence indicates that they [local people] knew quite well how to deal with a familiar hazard; it was the third in living memory …" (Waddell 1983, 33). However, the relief effort blocked the migratory pattern that was central to indigenous coping mechanisms. This effort undermined the confidence of people for whom that massive assistance had been intended when, in their minds, a disaster did not exist. Also, disaster relief actors sometimes view the afflicted from their own views of how things should be. They rarely seek the opinion of the disaster survivors and more often than not rely on their own cultural biases. Relief responses often impose remedies that are not appropriate for disaster-afflicted populations (Kent 1987).

In many cases, relief assistance may cause more damage than a given disaster itself in a long run. Receiving disaster relief after each event encourages people to settle in marginal or disaster-prone areas with the expectation that they will be compensated for their losses (Wisner 1993). For this reason, they are often reluctant to initiate mitigation measures and remedial actions to minimize losses from future extreme natural events (Haider et al. 1991).[11] Moreover, an influx of donated food, deemed necessary for saving disaster survivors from starvation, can lower market prices, discouraging local farmers from planting next season's crops and thus destabilizing the local agricultural economy (Smith 1992; Albala-Bertrand 1993; Tobin and Montz 1997; Bradshaw 2013).

Free food aid through disaster relief causes food prices drop, creating severe hardship for local farmers at a time when their employment has also been disrupted by the disaster (Alexander 2000). Low food prices often force farmers "to plant less than in the previous year due to lack of capital to purchase seeds and fertilizer" (Kent 1987, 13; also see Cuny 1983). Therefore, many disaster experts maintain that food aid should be reserved only as a last resort for survivors of extreme events (Paul 2011).

Smith (1992) claims that the flow of relief aid from developed to developing countries is often disaster-specific, highly unreliable, and may not reflect local needs. He argues that severe earthquakes and tropical cyclones that cause many fatalities generate a very high donor response. Alternatively, droughts and floods tend to produce relatively low responses, despite the large number of people who are adversely affected by these extreme events and need external support.[12]

Moreover, disaster aid "is often highly political and may even be used as a weapon by the powerful donor nations" (Smith 1992, 74). Thus, disaster relief is not always based on a humanitarian impulse but is often driven by politics and self-interest (Bradshaw 2013). Fink and Redaelli (2011) concluded that the countries in the Organization for Economic Cooperation and Development (OECD) provided emergency assistance out of political and strategies consideration as well as humanitarian concern after 270 emergencies worldwide in recent past.

International aid policies also are affected by historical links and geopolitical interests with the governments of disaster-affected countries. Foreign countries tend to favor allies, so more disaster aid generally flows to friendly nations. For examples, France quickly responds to disasters occurring in its former colonies, and the UK focuses upon members of the Commonwealth (Kent 1987).[13] Moreover, some disaster aid comes either with restrictions attached to its use or with a hidden purpose. For example, the donor transportation system must be used, or aid may be linked to other activities. This hidden agenda creates obligations on the part of aid-receiving countries. In such cases, the needs of disaster survivors are sometimes secondary to the political interests of donor and recipient countries (Tobin and Montz 1997).

During relief efforts for the 2004 Indian Ocean Tsunami (IOT), several Internet groups were formed in the USA by people of the affected regions. A close review of their postings shows that Western countries provide aid to create or expand markets and/or to achieve political and economic gains. According to these groups, Japan would naturally give disaster aid to repair damaged roads because the country has cars to sell. Denmark and Norway give aid and loans to improve the river transportation system, knowing that they will then be able to sell ships, launches, and ferries.[14]

Many of the criticisms of disaster aid center on the public distribution of disaster relief, which has often suffered severely because of widespread corruption and large-scale irregularities, most notably that such relief does not reach and benefit all disaster survivors equally (e.g., Smith 1992; Bolin and Stanford 1999). In essence, providing impartially distributed relief to disaster survivors has become less than common practice. Many claim that both private and public agencies systematically discriminate against marginal groups like the elderly, poor, permanently disabled, illiterate, women, those belonging to a lower caste, children, ethnic or religious minorities, and immigrants, promoting instead dominant-class interests (e.g., Blaikie et al. 1994; Bolin and Stanford 1999; Cutter 2003). Because of inequitable and biased distribution, disaster survivors belonging to marginalized groups benefit the least from disaster aid. This further exacerbates tensions between marginalized and non-marginalized groups after a disaster (Paul 2006).[15]

Apart from distribution issue, relief goods may not be culturally, religiously, or geographically appropriate. Further, relief items may not be provided in appropriate quantities, with deliveries of either too many or too few. Moreover, essential items are often not provided (Bradshaw 2013). Hazard studies provide ample evidence of useless and/or redundant aid disbursed to disaster survivors (see Chapter 5). Donating unnecessary and inappropriate goods creates problems with storage. These items occupy space that could be used to store more valuable disaster relief supplies. Generally, storage space for relief goods in developing countries is very limited. Moreover, transporting donated goods from abroad often costs more than value of the goods themselves. Donating food is also problematic because the safety and purity of donated canned foods cannot be assured. Often these goods are trapped in customs and never get to the disaster survivors at all (Paul 2006).

Other criticisms include difficulty in delivering disaster aid to remote areas and in war-torn countries, creating problems in timely distribution of emergency aid, which is crucial for disaster survivors because it saves lives and reduces suffering. Misuse, non-use, and mishandling of relief funds are problems associated with many disaster relief operations. Human rights violations are also concern in such operations. These issues are discussed in Chapter 3.

Need for Disaster Relief Aid

Despite so much criticism, there is a need for emergency relief aid for the survivors of extreme natural events. This need is acknowledged either explicitly or implicitly imbedded in many definitions of disasters. According to Randolph Kent (1987, 4), at the heart of most definitions of disaster "lies the assumption that the survival of directly affected groups will normally require assistance from outside that group – be it from neighbouring villages, central governments or from distant countries." This is also true for a considerable number of disasters definitions offered by hazard researchers after 1987. The Center for Research on the Epidemiology of Disasters (CRED) at the Catholic University of Louvain in Brussels, Belgium, provided an ideal example of such a definition of disaster.

The Center has developed the Emergency Events Database (EM-DAT) covering disasters in all countries from 1988 onward. This database is the primary reference among hazard researchers because its comprehensiveness and also because disaster agencies themselves use it. The primary goal of EM-DAT is to publish annually the frequency of disasters in each country of the world as well as other relevant information on those disasters. It defines a natural disaster as "a situation or event which overwhelms local capacity, necessitating a request to the national and international level for external assistance, or is recognized as such by a multilateral agency or by at least two sources, such as national, regional, or international assistance groups and the media" (Guha-Sapir et al. 2004, 16). For a disaster to be part of CRED's database, at least one of the following four criteria must be met:

- 10 or more people reported killed by the extreme event;
- at least 100 people reportedly affected (meaning they need immediate assistance during the post-disaster period);
- a request by the national government for international assistance; or
- the declaration of a state of emergency by the relevant government.

Among the four criteria, any appeal for international assistance from the country concerned usually takes precedence over all other criteria (Smith 1996). This definition of the CRED, which is used by the

U.N.'s International Strategy for Disaster Reduction (ISDR) and throughout the world, clearly suggests that disaster-affected communities cannot cope with losses caused by such destructive events without external support whether from within the country or beyond its borders (Picture This 2003). Donor countries and foreign aid agencies generally do not provide emergency assistance to survivors of disasters unless they receive a call for assistance. For this reason, governments of disaster-impacted countries appeal for emergency assistance immediately after the disaster.

The need for relief aid is also stressed in a definition of disaster provided by the United Nations Center for Human Settlement (UNCHS): "[A] natural disaster … which cause[s] severe losses to man and his environment (built and natural). These losses create suffering and chaos in the normal patterns of life, which lead to socioeconomic, cultural, and sometimes, political disruption. Such a situation requires outside intervention at international and national levels in addition to individual and communal responses" (UNCHS 1994). Outside intervention is also part of many other definitions of disaster. Spiegel (2005, 1915) defines a disaster as "a serious event that causes an ecological breakdown in the relationship between humans and their environment on a scale that requires extraordinary efforts to allow the community to cope, and often requires outside help or international aid."

Wisner et al. (2004, 50) claim that a "disaster occurs when a significant number of vulnerable people experience a hazard and suffer damage and/or disruption of their livelihood system in such a way that recovery is unlikely without external aid." They further emphasize the importance of outside intervention in the form of emergency relief to cope with the effects of disaster at the household level in their access model. This model states that the amount of access that people have to resources determines whether they can reduce their vulnerability and avoid disaster (Wisner et al. 2004). This implies that the distribution of disaster aid is heavily dependent upon the survivors' access to resources.

Several other definitions of natural disaster also note the need for non-localized response for the affected area to survive and recover from extreme events. Some of these definitions are provided in Table 1.3. One important point emerges from these definitions: natural disasters

Table 1.3 A select definitions of disaster

1. Gunn (1990, 374): "The result of a vast ecological breakdown in the relations between man and his environment, a serious and sudden event (or slow, as in drought) on such a scale that the stricken community needs extraordinary efforts to cope with it, often with outside help or international aid"
2. Disasters ".... cause damage, ecological disruption, loss of human life, deterioration of health and health services on a scale sufficient to warrant an extraordinary response from outside the affected community or area" (WHO cited in IASC 1994)
3. According to Cannon (1994, 29): "A disaster is an event associated with the impact of a natural hazard, which leads to increased mortality, illness and/or injury, and destroys or disrupts livelihoods, affecting the people or an area such that they (and/or outsiders) perceive it as being exceptional and requiring external assistance for recovery"
4. "A disaster is an unusually severe and/or extensive event that usually occurs unexpectedly and has such as severe impact on life and health of many people and/or causes considerable material damage and/or impairs or endangers the life of a large number of people for a long period of time to such an extent that resources and funding available at local or regional level cannot cope without outside help" (DKKV 2002, 216)
5. "A serious disruption of the functioning of a community or a society causing widespread human, material, economic or environmental losses which exceed the ability of the affected community/society to cope using its own resources" (UNISDR 2002, 24)
6. Disaster is "The result of a vast ecological breakdown in the relations between man and his environment, a serious and sudden event (or slow, as in drought) on such a scale that the stricken community need extraordinary efforts to cope with it, often with outside help or international aid" (*Journal of Prehospital and Disaster Medicine* 2004)
7. Disaster is "A serious disruption of the functioning of society, causing widespread human, material or environmental losses, which exceed the ability of affected society to cope using only its own resources" (EEA 2005)
8. NSTC (1996, 17): "A serious disruption of the functioning of a community or a society causing widespread human, material, economic or environmental losses which exceed the ability of the affected community or society to cope using its own resources"

Source Prepared by the author

invariably cause extensive damage and generally impose a severe economic burden on the survivors and affected communities in both developed and developing countries. This is particularly true for major disasters in developing countries that create demands typically exceeding resources available at the household and/or community levels. Therefore, survivors need outside support to help cope with losses, hardship, and distress caused by disasters. The extent of outside assistance need depends largely on the magnitude of the event. For some disasters, domestic sources beyond the affected community suffice to meet survivors' emergency needs, while larger events require support from not just within the affected country but outside as well.

Disaster survivors in both developed and developing countries not only need, expect, or receive relief from outside sources to mitigate the effects of extreme events, they often become proactive within their communities, building networks and providing assistance to other survivors (Mustafa 2003). Disaster relief represents an immediate response that alleviates the suffering of survivors and saves their lives (Minear 2002). Ideally, it is apolitical and reactive, short-term duration, and focused on meeting immediate basic needs and enhancing the disaster recovery process. Relief activities typically include providing food, medicine, clean water, sanitation services, temporary housing, and security, as well as psychological and social support (Bradshaw 2013). Disaster relief is not designed to mitigate or reduce the risk of future disasters nor reduce the long-term vulnerability of disaster survivors.

Disaster aid does not and cannot compensate for all losses caused by a disaster; it is a short-term (partial) damage reduction measure (Tobin and Montz 1997). It helps smooth the transition from necessary and immediate relief to long-term rehabilitation and redevelopment (Smith and Ward 1998). A smooth transition allows survivors to quickly return to pre-disaster life so they do not have to depend on any type of relief (Paul 2011). Rapid mobilization of relief supplies not only helps meet the immediate needs of the survivors; its effectiveness seems to be associated with the mental health of disaster survivors. For example, effectiveness of relief for Katrina was so poor that nearly 30% of all survivors 18 years or older suffered from at least some degree of post-traumatic stress disorder (PTSD) (McLaughlin et al. 2011). Bradshaw (2013, 100) concludes that "short-term relief projects may have longer-term implications for the material and emotional well being of those receiving the funding and those denied the funding."

Note that the impact of a disaster, particularly in developing countries, is generally influenced by the effectiveness of relief and emergency operations. With rising economic inequality, vulnerable populations must get support after a disaster. However, relief distribution alone is not enough to enable survivors to fully cope with the impact of disaster, although it is the first and most essential step toward disaster resiliency and recovery (Parvin and Shaw 2013).

Objectives of This Book

Providing aid to those suffering from a disaster is not a simple task; it is, rather, full of complexities. Many people and organizations are not necessarily familiar with these complexities at least in part because post-disaster relief has received limited attention from hazard and disaster researchers. This is ironic given the proliferation of relief actors in recent decades (Paulson and Menjivar 2012). Adding to the lack of available information is the near absence of any contemporary book on disaster relief operations and the complexities associated with such operations. Frederick Cuny discussed disaster aid in his 1983 book, *Disasters and Development*, with considerable information on development as part of the disaster relief context. Randolph Kent's 1987 book, *Anatomy of Disaster Relief: The International Network in Action*, presents a very thoughtful structural analysis of the international relief network, tracing the roles of government, intergovernmental agencies, and private voluntary organizations in disaster relief efforts, since the end of World War II.

However, disaster relief has changed, and new challenges have emerged in disaster relief disbursement since these books were published. Further, a large number of mega-disasters occurred after these books were published. Each of these disasters provided unique challenges to those people and organizations who provided and distributed emergency aid; the lessons learned from these events are important.

In 2011, Sage published an *Encyclopedia of Disaster Relief*, which contains 425 entries in a two-volume set, with contributors from many academic disciplines (Penuel and Statler 2011). The following year,

Routledge published a *Handbook of Hazards and Disaster Risk Reduction* (Wisner et al. 2012). Both volumes cover responses to disasters around the world, from governments to NGOs, from charities to politics, from refugees to health, and from economics to international relations, covering both historical and contemporary disasters. Although these publications provide a wealth of information and varied perspectives on disaster response and relief, they lack full-length treatment of number of relevant topics.

Thus, the principal objectives of this book are to (1) evaluate the current state of disaster relief, (2) update relevant literature, and (3) provide a multiscale view of emergency relief operations over the past few decades.

Through a historical approach, this book will shed light on progress in disaster relief endeavors across the globe. It has interdisciplinary and international appeal. It provides a comprehensive analysis of disaster aid, including the changes that have occurred over the last several decades, challenges that emerged, and the perpetual problems along with possible solutions. Over the years, disaster relief operations have become very complex. Knowing the complexities of emergency aid distribution will not only help people, but also help the agencies, providing and distributing emergency aid locally and internationally to improve future activities across the globe (MacRae and Hodgkin 2011). This book will provide valuable lessons for development agencies and the international humanitarian community, including many NGOs, nonprofit aid organizations, and bilateral and multilateral agencies, whose numbers have increased remarkably in recent years.

This book is much needed for emergency managers, public administrators, planners, policy-makers, and other decision makers working in areas prone to natural hazards as well as post-disaster environments. They can use the information in deciding what should or should not be done during the relief phase of the disaster management cycle as well as help them develop and train humanitarian actors whose numbers and diversity have also increased manyfold over the last several decades. Both undergraduate and graduate students of introductory or upper-level courses on hazards and disasters will also find the book essential and beneficial.

Note that although international development agencies have worked to prevent disasters since 1990 when IDNDR started, annual relief operations across the globe have not abated. More costly and more deadly disasters occur because of social marginalization and rapid environmental degradation and urbanization. Additionally, these agencies have widely accepted disaster preparedness and mitigation as integral components in the overall development process, so it comes as no surprise that many developing countries have given disaster management a very low-budgetary priority. These nations' resources tend to focus on more socially demanding interests like development of base infrastructure or on their military, not on projects associated with disaster preparedness and/or mitigation.

In the midst of all this, the number of people needing disaster relief has more than doubled in just over a decade. Several studies (e.g., Thomas and Kopczak 2005; Schulz and Blecken 2010; Nikbakhsh and Farahani 2011) claim that in the next fifty years, disasters will occur five times as often, mainly due to climate change, environmental degradation, and rapid urbanization. Additionally, the total number of recorded natural disasters has multiplied more than sixfold over the last 30 years (Hoyois et al. 2007; Schulz and Blecken 2010). This adds tremendous pressure on governments, humanitarian relief organizations, and aid agencies all over the world. Therefore, the disaster studies disciplines must improve how emergency relief can be provided, and this book will contribute to this end.

Recent development in social and networking media platforms, such as Facebook, Twitter, YouTube, and Instagram, as well as information technologies, such as smart phone, can also contribute to smooth distribution of disaster relief in future extreme events. To some authors (e.g., Tim et al. 2017), increased use of social media and information technology has resulted in a shift in disaster response paradigm from the traditional command-and-control response to digitally enabled disaster response. These new technologies have facilitated two-way communication in shortest possible time and provided a critical information source for the disaster response agencies and actors.[16]

The extensive use of social media was observed during the 2010 Haiti Earthquake as well as the 2012 Hurricane Sandy in the USA.[17]

The global response to the January 2010 earthquake in Haiti showed that individuals connected to the Internet are becoming increasingly central to humanitarian emergency responses. Haitians trapped under rubble as well as the Haitian Diaspora issued pleas for help or reported injuries using social media sites like Twitter, Flicker, Facebook, and blogs, and videos were posted on YouTube (HHI 2011). Haitian citizens sent over 30,000 text messages during the month following the earthquake, about 80,000 total messages posted (HHI 2011). Many of them published texts and photographs showing their personal experiences, and still more videos were posted on YouTube. In just 48 hours, the Red Cross received $8 million in donations directly due to texts, which exemplifies one benefit of the power of social media sites.[18]

Across the globe, thousands of ordinary citizens mobilized to aggregate, translate, and plot pleas of Haitian Earthquake survivors on maps and to organize technical efforts to support the disaster survivors. Many GIS specialists were able to quickly prepare necessary maps to facilitate emergency rescue and relief efforts using latest satellite imagery (Davey et al. 2013). This work, done through OpenStreetMap (OSM), became an essential element of the response (HHI 2011). These and other modern information technologies offer an alternative opportunity for first responders and disaster relief organizations to collect information and data about the disaster, survivors, and their needs. The International Federation of Red Cross and Red Crescent Societies (IFRC 2013, 20) termed this "humanitarian technology," which refers "to the use and new applications of technology to support efforts at improving access to and quality of prevention, mitigation, preparedness, response, recovery and rebuilding efforts."

Since the 2004 IOT, concerned citizens worldwide have become increasingly engaged in a variety of ways, from collecting donations through the Internet and sending in donations via SMS, to using shared networks to translate and map requests for assistance. These technologies help in not only collecting many donations in a relatively short time (Telford and Cosgrave 2007), but also challenging and changing the nature of humanitarian aid in emergencies. Mobile technology was also used in Aceh, Indonesia, after the 2004 IOT by survivors to comment on access to and the quality of relief work being performed by humanitarian agencies (HHI 2011).

After the 2010 Haiti Earthquake, a new group of stakeholders emerged: volunteer and technical communities (V&TCs). These stakeholders started to aggregate, geolocate, and prioritize incoming messages from various social media, providing geolocated reports of trapped individuals to urban search and rescue (USAR) teams on the ground. Overcoming many challenges, these stakeholder groups have proven to be a major player in disaster response and relief efforts by creating free damage and assessment maps and providing remote sensing images through OSM, Crisis Mapping, social media, and mobile devices (HHI 2011).

Chapter Arrangements

This chapter has begun by providing some salient features of disaster relief operations and a brief account of how global humanitarian assistance programs developed. Along with the major criticisms of provision of disaster aid distribution, the second section of this chapter also outlines the need for disaster assistance to survivors of extreme events. Finally, the objectives of this book are presented. In Chapter 2, information about how disaster aid from a plethora of private, public, humanitarian, and nonprofit organizations is presented. The chapter also highlights how the flow of aid is distributed from individuals (in both affected and non-affected areas/countries) to survivors. In addition, the problems and challenges associated with the flow of aid are presented. This chapter also covers other relevant information about providers and distributors of emergency aid, including participation of domestic and foreign military forces in disaster relief operations.

Crucial aspects of disaster relief distribution that ultimately determine the success of any single operation are presented in Chapter 3. This will provide background for understanding the strengths and weaknesses of disaster aid disbursement. The objective of Chapter 4 is to provide an overview of the nature and extent of emergency relief efforts undertaken in countries affected in the 2004 IOT, Hurricane Katrina, the Kashmir, and Haiti Earthquakes, and the Famine in South Sudan, along with some basic statistics on these disasters. This review is essential for improving future emergency relief operations. Additionally, each operation discussed in this chapter is fundamental for understanding the dynamics of the relief process.

Sending unnecessary and useless relief items to those that have suffered a disaster has been a perpetual problem in disaster relief operations. This spontaneous flow of supplies hampers and complicates emergency relief. This post-disaster convergence phenomenon is the subject of Chapter 5. There are three forms of convergence: personal (the influx of emergency response personnel and all types of individuals, including affiliated or non-affiliated local/national and foreign volunteers, and members of search, rescue, and medical teams); informational (the flow of information and communication); and material (movement of physical resources, like solicited or unsolicited donated goods, supplies, and/or equipment sent by all entities in response to a disaster). All three components of the convergence phenomenon and their effects on disaster relief activities are examined in this chapter. The final chapter (Chapter 6) outlines progress so far made in global disaster relief efforts during the last several decades, along with the challenges such efforts confronted. Based on these challenges, recommendations are made to improve future efforts.

Notes

1. Good Humanitarian Donorship (GHD) initiative established principles and good practices for donors of humanitarian aid. The 23 principles include four specifically on funding, two general principles related to funding, and one accountability principle pertaining to financial reporting (Good Humanitarian Donorship 2003). Donors endorsed these principles in a meeting held in Stockholm in 2003.
2. The Inter-Agency Standing Committee (IASC) is the primary mechanism for inter-agency coordination of humanitarian assistance. It is a unique forum involving key U.N. and non-U.N. humanitarian partners. The IASC was established in June 1992 in response to United Nations General Assembly Resolution 46/182 on strengthening humanitarian assistance.

3. The notion of a moral duty to collectively and/or individually respond to the suffering of humanity has existed for thousands of years, across many continents, and across many different cultures and faiths. For example, historical records show that as far back as 2000 BCE, an ancient Babylonian King urged his subjects to help widows, orphans, and the poor (Middleton and O'Keefe 1998).
4. Non-governmental organizations (NGOs) of developed countries are called NGOs of developed countries or international NGOs (INGOs) to distinguish them from NGOs of developing countries that began to emerge in the 1970s.
5. This led to a debate in the 1990s over the relationship between humanitarian assistance and development aid, a relationship called the relief-development continuum (Haider 2014). The continuum concept, which was ultimately rejected, sought to focus attention on the need to bridge funding gaps that existed at that time between emergency aid and development program.
6. Allegations regarding financial irregularities of NGOs in the handling of both development and disaster relief funds have started to surface since the late 1990s (Sharma and Zeller 1997; Mercer 1999).
7. For more details, see next chapter.
8. Despite this separation, humanitarianism is often linked to religion. For example, some Buddhist monks disbursed relief items only to Buddhists after Nepal hit by an earthquake in 2015 (Paul et al. 2017).
9. For the same reasons, floods seem to be frequent in Houston, Texas.
10. Just over two days, Hurricane Harvey dropped over 52 inches of rain in two days (Times-Picayune 2017).
11. No available empirical study shows to what extent the availability of disaster relief limits the adoption of hazard-reducing measures.
12. Despite his criticism of providing disaster relief, in his disaster impact pyramid, Smith (1992) acknowledges that such relief is provided and claims that the different effects of a major disaster can spread outward from a relatively small number of victims in the hazard zone to the global population through donations of aid and through the mass media.
13. This suggests that relief assistance often serves as apolitical and economic tool among donor nations. Moreover, such nations often respond to natural disasters because of domestic political pressure created either by politicians or by electronic media that increasingly bring

disaster images to the public and change public opinions in favor of providing assistance. For example, after the devastating tropical cyclone in East Pakistan (now Bangladesh) that occurred in 1970, the Government of Pakistan was very slow to respond to American offers of emergency assistance. The State Department warned the American Ambassador to Pakistan that the American public had begun to blame the US government for the slow response of the Pakistani government. President Nixon kept prodding the State Department to get the Government of Pakistan to respond so that he could keep Senator Kennedy off his back (Kent 1987). The same pattern has occurred in overseas disaster relief efforts in the UK, Japan, Germany, and elsewhere. Kent (1987, 78) further noted that "Political parties in one nation use humanitarian relief to demonstrate solidarity with similar ruling parties in a disaster-afflicted nation." Relief aid is also often seen as a means of satisfying particular pressure groups.

14. For example, the Dutch shipbuilding industry pressed hard for disaster assistance to Pakistan after its eastern province hit by a devastating cyclone in 1970. The industry wanted contracts to provide small river boasts to deliver relief materials to cyclone survivors (Kent 1987).
15. Research has provided little hard evidence that disaster aid has any net benefit.
16. Initial uses of social media were reported during the disaster relief efforts for the 2007 Southern California wildfire crisis, the 2008 Myanmar cyclone, and the 2008 Sichuan Earthquake in China. As indicated, social media-based relief efforts have been incorporated on a much larger scale since the 2010 Haiti Earthquake (Slagh 2010).
17. Unfortunately, social media were not used widely in Puerto Rico after it hit by Hurricane Maria on September 2017. This Category V storm made it completely isolated because the power grid went down and most cell towers were damaged.
18. Hurricane Sandy-impacted communities uploaded nearly 800,000 relevant photographs, and most of these photographs were geo-tagged with locations (Schifferes et al. 2014). These data provided a critical information source for all types of hurricane responders (DHS 2013).

References

Alam, J. 1988. Organizing the Rural Poor and Its Impact: The Experience of Selected Non-governmental and Governmental Organizations in Bangladesh. *Economic Bulletin for Asia and the Pacific* 39 (1): 33–42.

Albala-Bertrand, J.M. 1993. *The Political Economy of Large Natural Disasters, with Special Reference to Developing Countries*. New York: Oxford University Press.

Alexander, D. 2000. *Confronting Catastrophe: New Perspectives on Natural Disasters*. New York: Oxford University Press.

Alexander, D. 2006. Globalization of Disaster: Trends, Problems and Dilemmas. *Journal of International Affairs* 59 (2): 1–22.

Alexander, D. 2011. Chronology of Disaster Relief. In *Encyclopedia of Disaster Relief*, ed. K. B. Penuel and M. Statler, xxix–xxiv. Los Angeles: Sage.

Australian Red Cross. 2010. Spontaneous Volunteer Management: Resource Kit. http://www.fahcsia.gov.au/sites/default/files/documents?07_12012/spontaneous.pdf. Last Accessed 10 Aug 2017.

Ball, J.R. 2016. Louisiana Flood of 2016 Made Worse by Growth-Focused Policies. *The Times-Picayne*, Sept 21.

Barbarosoglu, G., L. Özdamar, and A. Cevik. 2002. An Interactive Approach for Hierarchical Analysis of Helicopter Logistics in Disaster Relief Operations. *European Journal of Operational Research* 140 (1): 118–133.

Barnett, M. 2011. *Empire of Humanity: A History of Humanitarianism*. Ithaca, NY: Cornell University Press.

Barnett, M., and P. Walker. 2015. Regime Change for Humanitarian Aid. *Foreign Affairs* 94 (4): 130–141.

Barraket, J., R. Keast, C. Newton, K. Walters, and E. James. 2013. Spontaneous Volunteering During Natural Disasters. Working Paper No. ACPNS 61, Brisbane, Queensland University of Technology.

Barrett, C.B., and D.G. Maxwell. 2005. *Food Aid After Fifty Years: Recasting Its Role*. London: Routledge.

Blaikie, P., T. Cannon, I. Davis, and B. Wisner. 1994. *At Risk: Natural Hazards, People's Vulnerability, and Disasters*. New York: Routledge.

Bolin, R., and L. Stanford. 1999. Constructing Vulnerability in the First World: The Northridge Earthquake in Southern California. In *The Angry Earth: Disaster in Anthropological Perspective*, ed. A. Oliver-Smith and S.M. Hoffman, 89–112. London: Routledge.

Booth, D. 1994. *Rethinking Societal Development: Theory, Research and Practice*. London: Longman.

Borton, J. 1993. Recent Trend in the International Relief System. *Disasters* 17: 187–201.

Bradshaw, S. 2013. *Gender, Development and Disasters.* Cheltenham, UK: Edward Elgar.

Bragg, C. 2015. Disaster Management and Multilaterial Humanitarian Aid: Parallelism vs. Combined Forces. In *The Humanitarian Challenges*, ed. F. Gibbons and H.-J. Heintze, 1–16. New York: Springer.

Cannon, T. 1994. Vulnerability Analysis and the Explanation of "Natural" Disasters. In *Development and Environment*, ed. A. Varley, 1–98. London: Wiley.

CDEM (Civil Defense Emergency Management). 2006. Spontaneous Volunteer Management Planning. http://www.civildefense.govt.nz/mem-website.nsf/Files/SpontaneousVolBPG306/Sfile/SpontaneousVolBPG306.pdf. Last Accessed 10 Aug 2017.

Chimni, B.S. 2000. Globalization, Humanitarianism and the Erosion of Refugee Protect. *Journal of Refugee Studies* 13 (3): 243–263.

Clay, E. 2003. Responding to Change: WFP and the Global Food Aid System. *Development Policy Review* 21 (5–6): 697–709.

Cnaan, R.A., F. Handy, and M. Wadsworth. 1996. Defining Who Is a Volunteer: Conceptual and Empirical Considerations. *Nonprofit Volunteer Sector Quarterly* 25: 364–383.

Cone, D.C., S.D. Scott, and S. Bogucki. 2003. Convergent Volunteerism. *Annals of Emergency Medicine* 41: 457–462.

Cuny, F. 1983. *Disasters and Development.* New York: Oxford University Press.

Cutter, S.L. 2003. The Vulnerability of Science and the Science of Vulnerability. *Annals Association of American Geographers* 93 (1): 1–12.

Davey, E., J. Borton, and M. Foley. 2013. A History of the Humanitarian System: Western Origins and Foundation. HPG Working Paper, Overseas Development Institute, London.

Davey, E., and K. Scriven. 2015. Humanitarian Aid in the Archives: Introduction. *Disasters* 39 (S2): S113–S128.

DHS (Department of Homeland Security). 2013. *Lessons Learned: Social Media and Hurricane Sandy*. Washington, D.C.: DHA.

Devereux, S. 2000. Famine in the Twentieth Century. IDS Working Paper 105, Brighton, Institute of Development Studies.

De Waal, A. 1997. *Famine Crimes: Politics and the Disaster Relief Industry in Africa*. Oxford: African Rights and James Currey.

DKKV (German Committee for Disaster Reduction). 2002. *Journalists Manual on DisasterManagement 2002*. Seventh Revised and Supplemented Edition. Berlin: DKKV.
Drabek, T.E., and D.A. McEntire. 2003. Emergent Phenomena and the Sociology of Disasters: Lessons Trends and Opportunities from the Research Literature. *Disaster Prevention and Management* 12 (2): 97–112.
Duffield, M., J. Macrae, and D. Curtis. 2001. Editorial: Politics and Humanitarian Aid. *Disasters* 25 (4): 269–274.
EEA (European Environment Agency). 2005. Multilingual Environmental Glossary. http://glossary.eea.eu.int/EEAGlossary. Accessed 20 Jan 2012.
Fink, G., and S. Redaelli. 2011. Determinants of International Emergency Aid—Humanitarian Need Only? *World Development* 39 (5): 741–757.
Fowler, A. 1990. Doing It Better? Where and How NGOs Have a Comparative Advantage in Facilitating Development. *AERDD Bulletin* 28: 12–16.
Fox, F. 2001. New Humanitarianism: Does It Provide a Moral Banner for the 21st Century? *Disasters* 25(4): 275–289.
Freeman, K. 1965. *If Any Man Build: The History of the Save the Children Fund*. London: Hodder and Stoughton.
GHD (Good Humanitarian Donorship). 2003. Principles and Good Practice of Humanitarian Donorship. 17 June. http://www.reliefweb.int/ghd/a%20 12%20Principles%20EN-GHD19.10.04RED.doc. Accessed 22 Dec 2016.
Greenough, P.R. 1980. Indian Famines and Peasant Victims: The Case of Bengal in 1943–44. *Modern Asian Studies* 14 (2): 205–235.
Guha-Sapir, D., D. Hargitt, and P. Hoyois. 2004. *Thirty Years of Natural Disasters: 1947–2003: The Numbers*. Louvain-la-Neuve and Belgium: Center for Research on the Epidemiology of Disasters and Presses Universitaries de Louvain.
Gunn, S.W.A. 1990. The Language of Disasters. *Prehospital and Disaster Medicine* 5 (4): 373–376.
Haider, H. 2014. Conflict: Topic Guide. Revised edition with B. Rohwerder. Birmingham: GSDRC, University of Birmingham.
Haider, R., A.A. Rahman, and S. Huq. 1991. *Cyclone '91: An Environmental and Perceptual Study*. Dhaka: Bangladesh Center for Advanced Studies.
Harvey, P., and J. Lind. 2005. *Dependency and Humanitarian Relief: A Critical Analysis. HPG Research Briefing*. Overseas Development Institute: London.
Hewitt, K. (ed.). 1997. *Regions of Risk: A Geographical Introduction to Disasters*. London: Longman.
HHI (Harvard Humanitarian Initiative). 2011. *Disaster Relief 2.0: The Future of Information Sharing in Humanitarian Emergencies*. Washington, DC: United Nations Foundation.

Hoyois, P., R. Below, J. Scheuren, and D. Guha. 2007. *Annual Disasters Statistical Review: Numbers and Trends.* Louvain: Université Catholique de Louvain and CRED.

IASC (Inter-Agency Standing Committee). 1994. Working Group 16th Meeting. Available at https://interagencystandingcommittee.org/system/files/legacy_files/WG16_4.pdf. Accessed 13 Oct 2016.

IFRC (International Federation of Red Cross and Red Crescent Societies). 1997. *World Disaster Report.* Geneva: IFRC.

IFRC (International Federation of Red Cross and Red Crescent Societies). 2013. *World Disasters Report: Focus on Technology and the Future of Humanitarian Action.* Geneva: IFRC.

Islam, R., and G. Walkerden. 2015. How do Links Between Households and NGOs Promote Disaster Resilience and Recovery? A Case Study of Linking Social Networks on the Bangladesh Coast. *Natural Hazards* 78: 1707–1727.

Journal of Prehospital and Disaster Medicine. 2004. Glossary of Terms. http://pdm.medicine.wisc.edu/vocab.htm. Accessed 11 Jan 2011.

Kent, R.C. 1987. *Anatomy of Disaster Relief: The International Network in Action.* London: Pinter Publishers.

Kibreab, G. 1987. *Refugees and Development in Africa: The Case of Eritrea.* Trenton, NJ: Red Sea Press.

Lewis, D.T. 1992. *Catalysts for Change? NGOs, Agricultural Technology and the State of Bangladesh.* London: Overseas Development Institute.

MacRae, G., and D. Hodgkin. 2011. Half Full or Half Empty? Shelter After the Jogjakarta Earthquake. *Disasters* 35 (1): 243–267.

Margesson, R. 2005. *Indian Ocean Earthquake and Tsunami: Humanitarian Assistance and Relief Operations.* Washington, DC: Congressional Research Service.

McLaughlin, K.A., P. Berglund, M.J. Gruber, R.C. Kessler, N.A. Sampson, and A.M. Zaslavsky. 2011. Recovery from PTSD Following Hurricane Katrina. *Depression and Anxiety* 28: 439–446.

Mercer, C. 1999. Reconceptualizing State-Society Relations in Tanzania: Are NGOs 'Making a Difference'? *Area* 31: 247–258.

Middleton, N., and P. O'Keefe. 1998. *Disaster and Development: The Politics of Humanitarian Aid.* London: Pluto Press.

Minear, L. 2002. *The Humanitarian Enterprise: Dilemmas and Discoveries.* Bloomfield, CT: Kumarian Press.

Mustafa, D. 2003. Reinforcing Vulnerability? Disaster Relief, Recovery, and Responses to the 2001 Flood in Rawalpindi, Pakistan. *Environmental Hazards* 5 (1): 94–105.

NSTC (National Science and Technology Council). 1996. *Natural Disaster Reduction: A Plan for the Future*. Washington, DC: U.S. Government Printing Office.

New World Encyclopedia. 2017. Disaster Relief. http://www.newworldencyclopedia.org/entry/Disaster_relief. Last Accessed 2 Sept 2017.

Nikbakhsh, E., and Z. Farahani. 2011. *Logistics Operations and Management*. New York: Elsevier.

OCHA (Office for the Coordination of Humanitarian Affairs) and Vodafone Foundation. 2011. Disaster Relief 2.0: The Future of Information Sharing in Humanitarian Emergencies. New York: UN. Available online www.unocha.org/top-stories/all-stories/disaster-relief-20-future-information-sharing-humanitarian-emergencies.

Oloruntoba, R., and R. Gray. 2006. Humanitarian Aid: An Agile Supply Chain? *Emerald* 11 (2): 115–120.

Parvin, G.A., and R. Shaw. 2013. Microfinance Institutions and a Coastal Community's Disaster Risk Reduction, Response, and Recovery Process: A Case Study of Hatiya, Bangladesh. *Disasters* 37 (1): 165–184.

Paul, B.K. 1998. Coping with the 1996 Tornado in Tangail, Bangladesh: An Analysis of Field Data. *The Professional Geographer* 50 (3): 287–301.

Paul, B.K. 2003. Relief Assistance to 1998 Flood Victims: A Comparison of the Performance of the Government and NGOs. *The Geographical Journal* 169 (1): 75–89.

Paul, B.K. 2006. Disaster Relief Efforts: An Update. *Progress in Development Studies* 6 (3): 211–223.

Paul, B.K. 2011. *Environmental Hazards and Disasters: Contexts, Perspectives and Management*. Hoboken, NJ: Wiley-Blackwell.

Paul, B.K., B. Acharya, and K. Ghimire. 2017. Effectiveness of Earthquakes Relief Efforts in Nepal: Opinions of the Survivors. *Natural Hazards* 85: 1169–1188.

Paulson, N., and C. Menjivar. 2012. Religion, the State and Disaster Relief in the United States and India. *International Journal of Sociology and Social policy* 32 (3/4): 179–196.

Penuel, K.B., and M. Statler (eds.). 2011. *Encyclopedia of Disaster Relief*. Newbury Park: Sage.

Picture This. 2003. Hit by Disasters. *Finance and Development* 40(3): 40–41.

Platt, R. (ed.). 2000. *Disasters and Democracy: The Politics of Extreme Natural Events*. Washington, DC: Island Press.

Raschky, P.A., and M. Schwindt. 2010. On the Channel and Type of Aid: The Case of International Disaster Assistance. Discussion Paper 06/10, Department of Economic, Monash University, Melbourne.

Rashid, H., and B.K. Paul. 2014. *Climate Change in Bangladesh: Confronting Impending Disasters*. New York: Lexington Books.

Schifferes, S., N. Newman, N. Thurman, D. Comey, A. Goker, and C. Martin. 2014. Identifying and Varifying News Through Social Media: Developing a User-Centered Tool for Professional Journalists. *Digital Journalism* 2: 406–418.

Schulz, S.F., and A. Blecken. 2010. Horizontal Cooperation in Disaster Relief Logistics. *International Journal of Physical Distribution and Logistics Management* 40 (8/9): 636–656.

Sharma, M., and M. Zeller. 1997. Repayment Performance in Group-Based Credit Programs in Bangladesh: An Empirical Analysis. *World Development* 25 (10): 1731–1742.

Slagh, C.L. 2010. Managing Chaos: 140 Characters at a Time: How the Usage of Social Media in the 2010 Haiti Crisis Enhanced Disaster Relief. Master's Thesis, Georgetown University, Washington, DC.

Slim, H., and J. Mitchell. 1992. Towards Community Managed Relief: A Case Study from Southern Sudan. *Disasters* 14 (3): 265–269.

Smith, K. 1992. *Environmental Hazards: Assessing Risk and Reducing Disaster*. London: Routledge.

Smith, K. 1996. Natural Disasters: Definitions. *Databases and Dilemmas. Geography Review* 10 (1): 9–12.

Smith, K., and R. Ward. 1998. *Physical Processes and Human Impacts*. New York: Wiley.

Solis, P. 1994. The Relief-Development Continuum: Some Notes on Rethinking Assistance for Civilian Victims of Conflict. *Journal of International Affairs* 47: 451–471.

Spiegel, P.B. 2005. Differences in World Responses to Natural Disasters and Complex Emergencies. *Journal of the American Medical Association* 293: 1915–1918.

Streeten, P. 1999. Globalization and Its Impact on Development Co-operation. *Development* 42 (3): 9–15.

Susman, P., P. O'Keefe, and B. Wisner. 1983. Global Disasters: A Radical Interpretation. In *Interpretations of Calamities*, ed. K. Hewitt, 274–276. London: Allen & Unwin.

Telford, J., and J. Cosgrave. 2007. The International Humanitarian System and the 2004 Indian Ocean Earthquake and Tsunamis. *Disasters* 31 (1): 1–28.

Therien, J.-P., and C. Lloyd. 2000. Development Assistance on the Brink. *Third World Quarterly* 21 (4): 21–38.

Thomas, A.S., and L.R. Kopczak. 2005. *From Logistics to Supply Chain Management: The Path Forward in the Humanitarian Sector.* Fritz Institute.

Thompson, J.B. 1990. *Ideology and Modern Culture.* Cambridge: Polity Press.

Tim, Y., S.L. Pan, P. Ractham, and L. Kaewkitipong. 2017. Digitally Enabled Disaster Response: The Emergence of Social Media as Boundary Objects in a Flooding Disaster. *Information Systems Journal* 27: 197–232.

Times-Picayune. 2017. Relatively Low Harvey Death Toll Is 'Astounding' to Experts. http://www.nola.com/hurricane/index.ssf/2017/09/harvey_death_toll_relatively_1.html. Last Accessed 2 Oct 2017.

Tobin, G.A., and B.E. Montz. 1997. *Natural Hazards: Explanation and Integration.* New York: The Guilford Press.

Townsend, J.G., G. Porter, and E. Mawdsley. 2004. Creating Spaces of Resistance: Development NGOs and Their Clients in China, India and Mexico. *Antipode* 36: 871–889.

Trivunovic, M., J. Johnson, and H. Mathisen. 2011. *Developing an NGO Corruption Risk Management System: Considerations for Donors.* Bergen, Norway: U4 Ant-Corruption Resource Center.

U.N. (United Nations). 1991. Strengthening of the Coordination of Humanitarian Emergency Assistance of the United Nations. www.un.org/documents/ga/res/46/a46r182.htm. Last Accessed 30 June 2017.

UNCHS (United Nations Center for Human Settlements). 1994. Sustainable Human Settlements in an Urbanizing World, Including Issues Related to Land Policies and Mitigation of Natural Disasters. 15th Session of the Commission on Human Settlement. Unpublished Draft Theme Paper.

UNDP (United Nations Development Program). 1999. *Human Development Report.* New York: UNDP.

UNDP (United Nations Development Program). 2004. *Reducing Disaster Risk: A Challenge for Development: A Global Report.* New York: UNDP.

UNISDR (United Nations, International Strategy for Disaster Reduction). 2002. *Living with Risk: A Global Review of Disaster Reduction Initiatives (preliminary version).* Geneva: UN ISDR.

Waddell, E. 1983. Coping with Frosts, Governments and Disaster Experts: Some Reflections Based on a New Guinea Experience and a Perusal of the Literature. In *Interpretations of Calamity from the Viewpoint of Human Ecology*, ed. K. Hewitt, 33–43. London: Allen & Unwin.

Walker, P., and D.G. Maxwell. 2009. *Shaping the Humanitarian World.* London: Routledge.

Walker, P., and K. Pepper. 2007. *Follow the Money: A Review and Analysis of the State of Humanitarian Funding*. Medford, MA: Feinstein International Center, Tufts University.

Washington County Emergency Operation Plan. 2011. Functional Annex—Emergent Volunteer Management. Portland, Oregon. http://www.ocem.org/Plans/FA-J_EmergentVolunteerManagement_Approved%206_3_11.pdf. Last Accessed 10 Aug 2017.

Watts, M. 1993. Development I: Power, Knowledge, Discursive Practice. *Progress in Human Geography* 17 (2): 257–272.

White, S.C. 1991. *An Evaluation of NGO Effectiveness in Raising the Economic Status of the Rural Poo: Bangladesh Country Study*. London: ODA.

Whittaker, J., B. McLennan, and J. Handmer. 2015. A Review of Informal Volunteerism in Emergencies and Disasters: Definition, Opportunities and Challenges. *International Journal of Disaster Risk Reduction* 13: 358–368.

WHO (World Health Organization). 2007. *Mass Casualty Management System: Strategies and Guidelines for Building Health Sector Capacity*. Geneva: WHO.

Wilson, J., and M. Musick. 1997. Who Cares? Toward an Integrated Theory of Volunteer Work. *American Sociological Review* 62 (5): 694–713.

Wisner, B. 1993. Disaster Vulnerability: Scale, Power, and Daily Life. *GeoJournal* 30: 127–140.

Wisner, B., P. Blaikie, T. Cannon, and J. Davis. 2004. *At Risk: Natural Hazards, People's Vulnerability and Disasters*. London: Routledge.

Wisner, B., J.C. Gaillard, and I. Kelman. 2012. *Handbook of Hazards and Disaster Risk Reduction*. London: Routledge.

World Bank. 2000. *World Development Report, 2000/2001: Attacking Poverty*. New York: Oxford University Press.

Zetter, R. 2003. *Shelter Provision and Settlement Policies for Refugees: A State of the Art Review*. Uppsala, Sweden: The Nordic Africa Institute, Swedish International Development Agency.

2

Channeling Disaster Aid: Process and Problems

This chapter includes essential topics associated with emergency relief operations, specifically those representing the early part of response stage/phase of disaster management cycle. This cycle has four overlapping temporal phases, two of which (mitigation and preparedness) transpire before a disaster occurs, while the last two (response and recovery) take place after an event (Alexander 2000; Coppola 2007) (Fig. 2.1). The cycle illustrates the continuous process by which many stakeholders plan for and reduce the consequences of disasters, assure prompt assistance to survivors immediately following a disaster, and take steps to achieve rapid and effective disaster recovery. Appropriate actions at all phases in the cycle lead to better preparedness and warnings, reduce potential fatalities, injuries, property losses, and prevent disasters in the next repetition of the cycle. The four stages of the disaster management cycle are not absolutely different from each other with considerable overlap among these stages/phases. The phases are useful categories that have helped organize planning, activities, research, and policy for hazard or disaster management and prevention.

Disaster response refers to all activities taken during and immediately following an extreme event, including efforts to save lives and to

B. K. Paul, *Disaster Relief Aid*,
https://doi.org/10.1007/978-3-319-77282-0_2

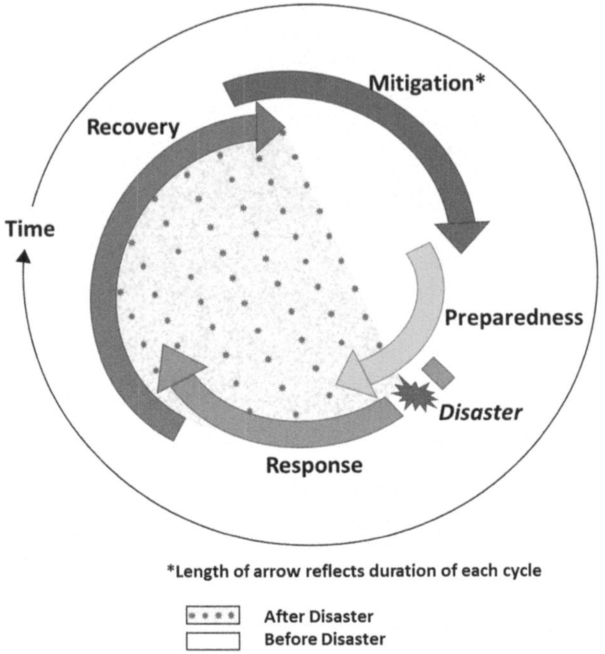

Fig. 2.1 Disaster management cycle (*Source* Modified after Alexander (2000, p. 3))

prevent further property damage. To Chou et al. (2014, 997), disaster response takes place "when a disaster is in progress and involves addressing immediate and short-term needs, including evacuation and rescue management." Ideally, disaster response involves putting already established disaster preparedness plans into motion. The main goal of the response stage is to provide for the immediate needs of those who are affected by the disaster and prepare for them for recovery. Although this stage of disaster management cycle includes many activities (e.g., search and rescue operations; emergency medical care for the injured; mental health counseling; proper identification and disposition of dead bodies; debris removal; provision of temporary shelter, water, sanitation, and food; repair of utilities and key infrastructure; security for survivors and protection of property; and closure of roads and bridges), the response stage starts with rescue operations initially by local residents and later by trained and skilled domestic as well as foreign search and rescue (SAR) teams. Rescue operations usually last for the first 48–72 hours after a disaster.

Relief operations follow immediately after rescue operations and, depending on the severity of the disaster and resources of the government of affected country, may last from three months to a year with a peak in the middle (Paul 2011). In some disasters, these two operations overlap. Like rescue operations, relief operations start with local people who are subsequently joined by domestic and foreign individuals and teams within 24 hours of the disaster or later. Thus, a plethora of private, public, humanitarian, and nonprofit organizations, and individuals from both affected and non-affected areas/countries are involved in disaster relief operations (Coppola 2007; Paul 2011). The focus of such operations is to provide basic necessities to disaster survivors. Detailed assessment of human and other losses is also usually carried out during the relief phase, which helps in optimal allocation of resources (see Post-disaster Needs Assessment (PDNA) subsection of this chapter).

The recovery stage of the disaster management cycle begins either at the end of or after the response stage, and the primary goal of this stage is to return livelihoods of survivors as much as possible to pre-disaster levels (FEMA 2000). It encompasses repairing and reconstructing houses, commercial establishments and public buildings, and infrastructure (Paul 2011). Depending on the level of economic development, the recovery stage lasts for several years to more than a decade (Middleton and O'Keefe 1998). Following recovery is the mitigation stage, which refers to policies and activities that eliminate or reduce the probability of disaster occurrence or reduce the effects of unavoidable disasters (Fig. 2.1). Mitigation measures are disaster specific and include building codes and building standards, land-use planning and regulations, construction of public shelters, safe rooms, flood walls, dikes and levees, and public education and installing hazard warnings. These measures generally are in place before, during, or after a disaster (Mileti 1999).

Immediately before the response stage is the preparedness stage, which refers to the degree of alertness and readiness of an individual, household, or community for an impending disaster (Fig. 2.1). Preparedness activities should help save lives and minimize property damage; these activities include formulating, testing, and practicing disaster plans, hazard warnings and evacuation, communicating disaster risk to the public as well as what to do to reduce risk, conducting emergency response drills, and providing disaster training for emergency responders and the public (Tierney et al. 2001). The preparedness stage is shorter than the response stage, indeed, the shortest duration of all four stages of the disaster management cycle.

Since this book emphasizes disaster relief operations, most topics in the book will cover activities associated with the relief phase of the response stage of the disaster management cycle. Accordingly, this chapter covers relevant and necessary information in chronological order of how disaster aid and emergency supplies flow from non-affected areas/ countries to survivors of major extreme events, who provides aid and supplies to the survivors, what problems and challenges are associated with disaster aid flows, and how to improve this flow. The information presented in this chapter helps further explain donation supply chains, which will provide better insight into how to more effectively manage disaster donations and emergency aid flow.

Flash Appeal

Anatomy of Disaster Appeal

The Flash Appeal is a tool for structuring a coordinated strategic response for the first three to six months after a major natural or man-made disaster (ReliefWeb 2008). It is a process for identifying emergency needs and inventorying relief and early recovery projects (Batha 2005). The U.N. Resident Coordinator or Humanitarian Coordinator (RC/HC) triggers the process in consultation with other U.N. agencies, large nongovernmental organizations (NGOs) like the International Committee of the Red Cross (ICRC), the government of the affected country, and other stakeholders. Closely associated with the Flash Appeal is the Consolidated Appeal Process (CAP), established in 2006 as a planning and programing tool to allow "aid organizations together to jointly plan, coordinate, implement and monitor their response to natural disasters and complex emergencies, and thus appeal for funds justifiably, cohesively and collectively instead of competitively" (OCHA 2012, 1). The CAP fills funding gaps, prevents funding overlaps, ensures that projects complement each other, and takes a strategically coherent approach to major disasters.[1] A Flash Appeal is generally issued within one week of the onset of a natural disaster that requires a coordinated response beyond the capacity of the government of the affected country to respond to, even in combination with any single U.N. agency. Table 2.1 shows the ideal timeline for the Flash Appeal procedure.

The Flash Appeal or appeal for international assistance can be applied in a range of contexts, even complex humanitarian emergencies caused by political, racial, or religious action. It is not generally issued for slow-onset disasters because in the case of such slow-onset disasters, it is difficult to provide evidence if the situation becomes suddenly worse. Flash appeals are revised as new information comes in. For example, the U.N. OCHA drafted a Flash Appeal on January 6, 2005, in the amount of $1.1 billion, for emergency aid for the December 24, 2004, Indian

Table 2.1 Ideal timeline for Flash Appeal procedure[a]

Day	Task
Day 1	The UN RC/HC initiates Flash Appeal
Day 2	Begins needs assessment
Day 3	Drafts of general sections of appeal document by the RC/HC
Day 4	RC/HC and country Inter-Agency Standing Committee (IASC) team approve final field draft of the appeal and sends it to OCHA CAP Section
Day 5	The CAP Section sends the final draft to IASC agency headquarters for 24-hour review
Day 6	After receiving from IASC, the CAP Section makes necessary revisions, including format checking of the document. The CAP Section then registers the appeal on the Financial Tracking Service, publishes the document online (ReliefWeb), and prints copies by 13h00. After printing the appeal is officially announced either through a donor meeting or through a press release
Day 7	Appealing agencies start receiving funds from donors and update on funding received. Also, agencies inform whether any project change occurs

[a]This timeline may vary according to the situation
Source Prepared by the author based on IASC (2006, pp. 3–4)

Ocean Tsunami (IOT). This appeal was made to fund the work of some 40 U.N. agencies among them, the U.N. Children Fund (UNICEF), the World Food Program (WFP), and the U.N. Development Program (UNDP), and NGOs like Save the Children, Cooperative for Assistance and Relief Everywhere (CARE), Catholic Relief Service, and International Rescue Committee in addressing the survival and recovery needs for an estimated 5 million people affected by the disaster (Schreurs 2011). The initial Flash Appeal, which was requested before January 6, 2005, for the IOT, was for US$977 million. The final revised appeal in April 2005 called for $1086 billion (Schreurs 2011).

A more recent example of Flash Appeal revisions was for the April 25, 2015, Nepal Earthquake. The initial Flash Appeal was issued on April 29 by the U.N. and its humanitarian partners to international donors to provide $415 million to respond to the most urgent needs of those affected by the earthquake over the following three months. These needs included food, shelter, and water, sanitation and hygiene (WASH) (OCHA 2015a). The Flash Appeal reflects both the scale of the needs and the significant logistical challenges linked to providing an effective large-scale humanitarian response in hard-to-reach, mountainous areas. Under this appeal, which was jump-started with $15 million made available through the U.N. Central Emergency Response Fund (CERF), partners would provide emergency shelter to 500,000 people who were sleeping under open sky since the major earthquake on April 25, 2017. Emergency health services and medical supplies and facilities, as well as safe drinking water and sanitation facilities, were also urgently needed for up to 4.2 million people. Before the U.N. appeal, within the hours of the earthquake, the Government of Nepal made an official request for international assistance for the survivors of the event (GoN 2015).

Because of numerous aftershocks, particularly the 12 May major aftershock of 6.8 magnitudes, the U.N. revised its Flash Appeal on May 29 because of the major and minor aftershocks that caused further damage and increased the number of people in need of emergency assistance. The Flash Appeal called for $422 million to provide life-saving and protection needs and support resilience for 2.8 million people over five months (OCHA 2015b). The appeal covered all vulnerable groups, including women, children, people temporarily displaced by the earthquake, host communities, ethnic and indigenous groups, and other affected people. The appeal duration was extended beyond three months because of the effects of the monsoon would have on relief operations and to ensure those relief operations would be linked to the recovery and rehabilitation program of the Nepali government. It is worthwhile to mention that in response to Flash Appeal for Nepal, 78 foreign relief organizations proposed 183 projects with a total funding request of $422 million. Ironically, only eight Nepali organizations applied for funding with requests totaling only $3.5 million or 0.8% of the total (Troutman 2015).[2]

However, a Flash Appeal should take into account the actions and plans of entities not in the appeal, such as the government and the office of the International Federation of Red Cross and Red Crescent Society (IFRC) of the affected country. It should include priority projects from all key humanitarian organizations on the ground. Flash Appeals should be issued as a concise document based on available information and reasonable inference, focusing on urgent humanitarian needs. It is brief and simply written, so people can read it quickly, and donors can respond without delay. A Flash Appeal does not generally exceed 10 pages, including any charts and tables that explain issues wherever possible (IASC 2006).

Flash Appeals should not be used when the affected country can conduct emergency responses using its own resources and trained manpower. Thus, the appeal is generally issued when major disasters strike in developing countries. As noted, Flash Appeals might be hard to justify for slow-onset disasters. It should also not be used when the situation requires more recovery-related activities or activities that take longer than the normal six-month Flash Appeal period. The U.N. generally makes the Flash Appeal to the international community, and the permission of the government of disaster-affected country for such an appeal is not required.[3] As noted, the U.N. CERF provides the initial injection of funds for the most urgent life-saving projects in the Flash Appeal to cover the time lag between issuance of the appeal, and receiving commitments and funds from donors. CERF (a funding instrument) does not replace a Flash Appeal (a planning tool); they interact. Most emergencies need more than CERF can provide.

Customarily, NGOs do not issue Flash Appeals, but they do provide vital assistance and distribute emergency aid. During some disasters, however, NGOs will appeal for international assistance. For example, when Bangladesh experienced severe flooding in early August 2007 that affected about 10 million people of the country, the Bangladesh government appealed to its citizens to help people affected by the flood, but it did not appeal to the international community. Observing this, Oxfam did appeal to the international community for assistance (Paul 2011). Donor nations may also ask the government of a disaster-affected country to appeal for international assistance. Thus, a Flash Appeal for a disaster can be issued by the government of the affected country, other countries, and/or large NGOs. At least four entities issued Flash Appeals over four days after the 2010 Haiti Earthquake (Table 2.2).

Early recovery projects implemented within the time frame of the appeal are a part of humanitarian actions and therefore can and should be included in Flash Appeals. As noted, given changing situations and as better assessment information becomes available, the appeal can be revised within about a month of being launched (Paul 2011). Revisions should incorporate (1) more complete information, (2) improved and in-depth assessments, and (3) more clearly defined early recovery projects. Thus, a Flash Appeal is not a static document but open and flexible. Flash Appeal guidelines have three parts (IASC 2006). The first part describes the Flash Appeal concept with information on background, policy, and practice on key substantive issues like timeliness and deadlines as well as revising Flash Appeal. The second part provides technical guidelines for writing Flash Appeals and Flash Appeal revisions, and

Table 2.2 Entities issued appeals for the January 12, 2010, Haiti Earthquake

Date	Entity
12 January	Haiti Government made an appeal for international assistance
13 January	US ambassador to Haiti issued a disaster declaration
13 January	IFRC launched preliminary emergency appeal
15 January	The UN issued a Flash Appeal for emergency financial assistance in the amount of $575 million

Source Prepared by the author based on information provided by Margesson and Taft-Morales (2010, pp. 3, 14) and U.N. (2010, p. 1)

the final part lists further resources like selected guidance documents, thresholds for triggering Flash Appeals, examples of priorities, financial tracking tips, and guidelines for the U.N. OCHA, which drafts and leads Flash Appeals (IASC 2006).

A Flash Appeal is generally divided into categories, or clusters like food, shelter, water/sanitation, health, and education. As indicated, apart from the U.N., often relief agencies like IFRC and Oxfam make separate appeals. These agencies generally get their own appeals out much faster than the U.N. launching them within a day or two of a disaster and then revising them as necessary. The IFRC may even issue an appeal in advance of a disaster if forecasting provides advance warning for a specific area (Batha 2005).

Flow of Relief Aid and Pledge

Not all international relief actors can formally respond to a disaster unless the recipient government specifically requests help (Kent 1987). Such a request implies that the government cannot cope with a disaster without help from other countries; some disaster-affected countries are reluctant to admit this truth because of ideological and/or economic issues (Bradshaw 2013). The recipient government may reject offers of assistance from foreign governments for political reasons or accept international assistance only if it comes through the appropriate channels established by the recipient government. Foreign governments often use such a request to establish a foothold in the disaster-affected country expecting the government to lose political control once the doors are open to international actors (Bradshaw 2013). However, in most cases, recipient countries are more than willing to accept emergency relief items because without external aid, these countries cannot effectively respond to the crisis.

Donor governments are by far the most significant component of all international relief actors or networks. Donor governments generally use three basic channels to provide relief: (1) bilateral assistance between donor and disaster-affected governments; (2) multilateral assistance through intergovernmental organizations or international governmental organizations (IGOs); and (3) indirect assistance either to recipient governments or directly to the afflicted through NGOs (Kent 1987; Fuchs and Nils-Hendrik 2012). Bilateral aid offers the advantage of speed. But the donors must explain to legislative bodies and various governmental accounting offices how and where aid has been gone. This can be a problem if recipients are corrupt with little accountability for their action or where the complexity of a disaster makes accurate accountability very difficult. Foreign private donations are primarily channeled through foreign NGOs and charities.

Although IGOs are perceived as slow to respond in emergencies and of being more concerned with their own institutional interests than with the plight of the afflicted, donor governments and other international humanitarian actors often use them for several reasons. IGO representatives are present in most disaster-prone countries. For example, almost all countries have one or more United Nations agencies present, such as UNDP, UNICEF, FAO, and WFP.[4] Additionally, the World Bank has offices in almost all countries. These missions/agencies have established contacts within recipient governments. This facilitates relatively smooth disbursement of relief materials through IGOs to disaster survivors. Not only do IGOs have access to those government officials and departments within a host country, but they are also familiar with general infrastructure and resource conditions of the country and can offer a degree of relief expertise. Most IGOs have either in-house experts or available consultants who can assess disasters and determine appropriate relief needs. Finally, IGOs can off-load responsibility for monitoring and assessing the impact of aid (Kent 1987). By using IGOs as a channel, a donor government avoids the political implications of bilateral assistance. These organizations generally present themselves as politically neutral and only concerned with humanitarian issues. Similarly, the recipient governments regard IGO missions as relatively apolitical.[5]

When the donor governments disburse disaster aid through IGOs, they often earmark aid funds to be used solely for tents or medicine or specific commodities to be distributed to particular types of disaster survivors.[6] Either ways, the donor expects the IGO to comply with such restrictions and to report how these contributions have been used in the field. Earmarking, however, has several problems. For example, the actual needs of the disaster survivors may not correspond with what donor government perceives. This results in the flow of unnecessary relief items to disaster sites. Earmarked funds can lead to duplication of supplies, leading to convergence of materials in the site. Earmarking also creates a dilemma for the IGO itself; not accepting donor's contributions may alienate the donor, so in the future, the donor may not channel aid through that particular IGO. Additionally, earmarking means the IGO does not have control and cannot assess results (Kent 1987; Toyasaka and Wakolbinger 2014).

Donor governments often channel disaster funds through a multi-donor trust fund administered by the World Bank. OCHA can organize a trade fair where relief agencies display their needs in stalls and donor entities choose what they want to fund (Paul 2011). Although each disaster is unique, the donor governments often use all these channels, particularly in large-scale emergencies. The direct financial contributions of most IGOs to relief, however, are relatively small compared to bilateral aid.

A careful review of the amount of disaster assistance committed by humanitarian actors in the past few decades reveals a large gap between pledged amount and actual delivered amount in most instances. A pledge is a non-binding announcement of an intended contribution or allocation by the donor for a natural or man-made disaster (ReliefWeb 2008). The gap not only differs from one disaster to another, but also from one donor entity to another. For example, for the 2003 Bam Earthquake in Iran, donor nations actually delivered only $17.5 million of the $1.1 billion pledged (1.5%) (Powell 2004). In contrast, as of December 2016, donor nations provided 92% of the committed funds for the 2004 IOT (see Chapter 4, Table 4.1). The governments of non-affected countries provided 77% of the pledged amount, while IFRC and Islamic Development Bank provided 100% of their pledged amount.

Several reasons account for the gap between pledged funds and the amount actually released. Sometimes donor entities feel pressure to pledge more funds than they can actually provide, resulting in a large gap between pledge amount and actual amount donated for disaster survivors. Telford and Cosgrave (2007) called this a donor "beauty contest." Double count is often also responsible, particularly when some promised funding is diverted from existing aid. Competition between two or more donors may increase pledged amount, which happened with China and Japan after the 2004 IOT. Japan wanted to provide more funding to tsunami-affected Asian countries than China to show that Japan is economically more powerful than its traditional rival (Tamamoto 2005).

Donor countries also often need to increase their original pledged amount for several times. For example, the USA originally pledged only $15 million for the IOT, but in response to criticism from U.N. Under-Secretary-General for Humanitarian Affair Jan Egeland, the USA added another $20 million to its original pledge and then increased the pledge to $350 million. As the death toll increased and European contributions soared, the USA once again raised its pledge to $857 million (Wehrfrtiz and Cochrane 2005; Paul 2011). However, how much donor countries really increase a disaster pledge is often unknown because the pledge may replace previously pledged amount (Becerra et al. 2014).

Note that the pledged amount is for both relief and recovery; generally almost identical amounts are spent for both. However, most donors strongly favor one over the other, while some donor offer funds mainly for emergency needs. For example, approximately one-half of the pledges made for the IOT in January 2005 were for emergency assistance; one-quarter for recovery to be spent in 2005; and the remaining quarter for recovery between 2006 and 2010 (Telford and Cosgrave 2007). Often total pledge amount exceeds the estimated losses and damage. For example, the pledge of $10 billion for the 2004 IOT exceeded by at least one-third. However, the estimates of damage and loss do not necessarily include the full cost of either the initial relief operation, or the cost of temporary services and shelter while waiting for permanent replacements (Telford and Cosgrave 2007).

Cluster Approach

In 1991, the U.N. created the OCHA to improve coordination among the many organizations and stakeholders involved in relief operations for natural disasters and complex emergencies. Given the coordination failures both during the humanitarian response to the 2004 Darfur crisis in Sudan as well as disorderly response to the 2004 IOT relief efforts, OCHA felt the urgent need to improve relief response in the international humanitarian system (Humphries 2013).[7] Relief efforts for both these events were delayed and inadequate primarily due to the inability of involved organizations to mobilize their capacity and resources, not to mention the lack of information sharing among the participants (Gillmann 2010). With the primary goal of improving the effectiveness and timeliness of international humanitarian responses to emergencies, the OCHA introduced coordination using a Cluster Approach in 2005, which are formally implemented in 27 of the 29 countries with U.N. Emergency Relief Coordinators (Gillmann 2010).

Other goals of the Cluster Approach are to improve relief responses in the international humanitarian system by sharing information and minimizing duplication of services delivered by humanitarian actors. This approach ensures that various organizations are synchronized, working in harmony to achieve a common objective, making emergency response more effective, efficient, and accountable (Humphries 2013), thus facilitating partnerships among humanitarian actors and improving inter-agency complementarity by maximizing resources. (ReliefWeb 2008). Clusters are sectors of humanitarian actions (e.g., food security, health, and shelter), with the aim of filling gaps, ensuring adequate preparedness and response, and prioritizing available resources. Each sector is led by a large organization, preferably U.N. agencies like the High Commission for Refugees or the WFP, designated by the Inter-Agency Standing Committee (IASC). Each lead agency has clear responsibilities for coordination.

The Cluster Approach was first applied following the October 2005 earthquake in Pakistan, also called the Kashmir Earthquake. Nine clusters were established within 24 hours of the earthquake (Street and Parihar 2007). However, the number of clusters differs from one disaster to another. Ten clusters were set up for the February 2007 flood in Mozambique with the support of an OCHA humanitarian officer from the Humanitarian Reform Support Unit in Geneva. These 10 clusters are listed in Table 2.3. Between these two events, the Cluster Approach

Table 2.3 Ten clusters identified for the February 2007 Mozambique flood

Cluster name	Cluster lead	No. of cluster participants*
Logistics	World Food Program (WFP)	29
Food security	WFP/WHO	20
Telecommunications	WFP	5
Water, sanitation, and hygiene	UNICEF	12
Nutrition	UNICEF	11
Health	WHO	7
Education	Save the Children Alliance/UNICEF	8
Protection	Save the Children Alliance/UNICEF	17
Emergency shelter	Mozambican Red Cross/IFRC	8
Early recovery	UNDP	All parties

*A partial list of cluster participants was: IFRC, USAID, Oxfam, World Vision, Caritas, ActionAid, Concern International, Food for the Hungry International, German Agro Action, Care, UNAIDS, Africare, and Habitat for Humanity
Source Prepared by the author based on information provided by Actionaid (2007, p. 13)

was implemented in four major emergencies in Indonesia, Lebanon, Madagascar, and the Philippines (Actionaid 2007). Twelve clusters each were formed for the 2010 Haiti and 2015 Nepal Earthquakes. For the 12 clusters identified in the initial Flash Appeal for the 2015 Nepal Earthquakes, the total financial requirements were estimated to be $415 million from April to July 2015 to reach 8 million survivors with emergency assistance (OCHA 2015a). The initial appeal was revised on May 29, 2017, and increased the amount to $422 million. The appeal duration was extended to September to take into the account the effects of the monsoon on the emergency phase (OCHA 2015b). Figure 2.2 illustrates financial requirements per cluster for both initial appeal and revised appeal. Each cluster listed the number of priority actions for both Flash Appeals. For example, the initial appeal had five priorities includingwater, sanitation, and hygiene (WASH) cluster, which were:

- Provide safe water for the most vulnerable internally displaced peoples (IDPs) and for institutional facilities,
- Provide temporary latrines and bathing spaces for the most vulnerable IDPs and for institutional facilities,

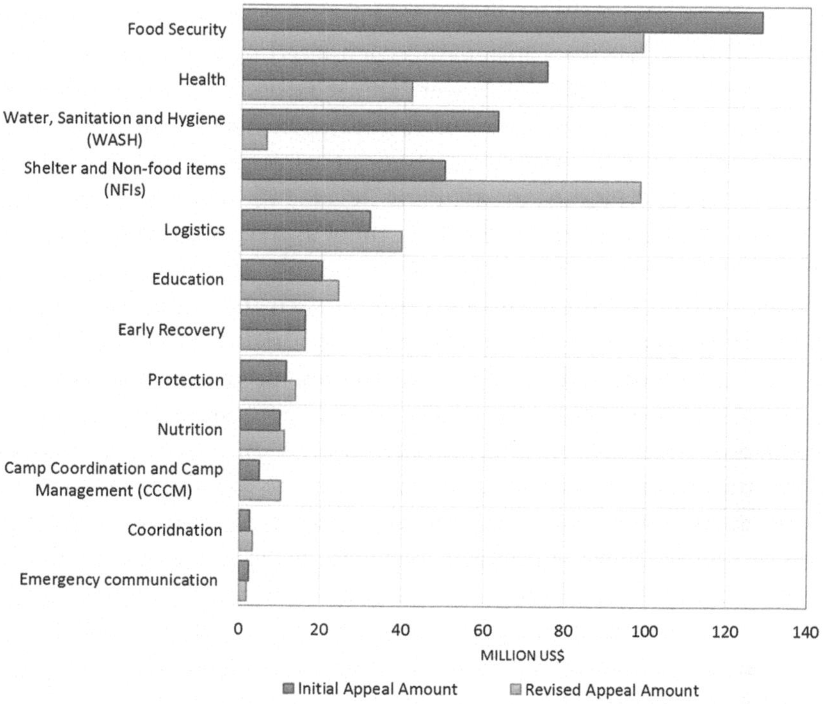

Fig. 2.2 Financial requirements per cluster for both initial and revised appeals for the 2015 Nepal Earthquakes (*Sources* Adapted from OCHA (2015a, p. 7) and OCHA (2015b, p. 18))

- Distribute hygiene kits to families in need,
- Implement hygiene promotion campaigns to reduce the risk of waterborne disease outbreaks, and
- Provide limited collection of solid waste in IDP camps and conflict-affected areas (www.unocha.org/nepal).

Although humanitarian disaster relief efforts have improved after the commissioning of the Cluster Approach, a few challenges remain. First, because huge numbers of individuals and organizations converge at the site of major disaster, the existing coordination mechanisms outlined by the approach appear to have reached their limits. This was particularly evident during the 2010 Haiti Earthquake relief operation. Humphries (2013) claimed that the Cluster Approach faced problems because cluster coordination had high turnover rates, coordinators lacked adequate training and experience, and cluster lead agencies lacked impartiality. Second, cluster coordination is labor intensive and time-consuming. Moreover, many NGOs, particularly local and regional NGOs, are still uncertain about how effective the Cluster Approach has been (Humphries 2013).

Some NGOs feel the approach conflicts with the humanitarian principles of independence, neutrality, and impartiality, which are crucial in reaching the affected people. Finally, the Cluster Approach does not have enough mechanisms in place to offer accountability to affected populations (Humphries 2013). The Cluster Approach offers many opportunities for improvement, so it is a positive change in coordinating relief. As people and organizations participate in future relief operations using the Cluster Approach, they gain experience, which will help improve coordination going forward, since large-scale disasters occur almost every year, so gaining experience should not take much time. In the meantime, those in charge of clusters approach should review their work and revise the system using their past experience.

Good Humanitarian Donorship (GHD)

The GHD initiative is an informal donor forum and network that collectively facilitates advancing GHD principles and good practices. This initiative was launched in 2003 at an international meeting attended by representatives of donor governments, U.N. agencies, the International Red Cross and Red Crescent Movement, and host of other organizations involved in humanitarian actions in different countries. In 2005, the Development Assistance Committee of the Organization for Economic Cooperation and Development (OECD/DAC) adopted it. Attendees of the 2003 meeting established 23 principles and good practices for donors of humanitarian aid (Telford and Cosgrave 2007). Each year, two donor countries cochair the initiative and convene ad hoc meetings of GHD donor representatives to evaluate the progress of pilot approaches and discuss implementing approaches in multilateral forums. Their Web site posts regular updates on progress and provides a comprehensive list of resources (ODI 2006).

Of the 23 principles included in the GHD, three give the objectives and definition of humanitarian action, seven state general principles, four elaborate on funding, six provide standards and enhance implementation, and the remaining three concern learning and accountability principles pertaining to financial reporting (Good Humanitarian Donorship 2003). The five main strands of the GHD implementation plan are as follows: (1) donors must identify at least one crisis subject to a CAP to which the principles and good practice defined by GHD will be applied; (2) donors must invite the OECD/DAC to consider ways to strengthen peer review to include humanitarian action; (3) donors must explore how humanitarian organizations can harmonize reporting requirements and balance management demands; (4) donors (with the U.N. and OECD/DAC) must agree on a comprehensive common definition of official humanitarian assistance (OHA) for reporting and statistical purposes; and (5) donors must encourage the wider use of GHD's principles and good practice, and invite all interested donors to participate in following up on the implementation plan (Good Humanitarian Donorship Initiative, n.d.).

Although not adopted by all donors, the initiative is a major step forward in promoting more transparency and accountability as well as improved coordination among donors. However, donor governments have had difficulty incorporating the GHD principles in their policies and practices, as shown in many evaluations like the major evaluation after the IOT, and of internally displaced persons (IDPs) in Sudan and Sri Lanka (Collinson et al. 2009). Moreover, the GHD lacks specific targets and benchmarks, making progress in implementation at both the global level and the level of individual donor governments difficult to determine (Schaar 2007).

Post-disaster Needs Assessment (PDNA)

The PDNA is a powerful tool for assessing the damage and losses caused by natural disasters and then developing a comprehensive recovery plan. PDNA is effort product of the joint collaboration among the European

Union (EU), the World Bank, the United Nations Development Group, and other international agencies, such as the Asian Development Bank (ADB) and IFRC. Immediately after a developing country experiences a major natural disaster, the government of the affected country requests international aid communities and development partners to quickly prepare a report on disaster damage, loss, and needs for recovery and reconstruction. Without such request, the appropriate authority will not prepare a PDNA. Note that the developing countries affected by disaster do not necessarily make this request for all natural disasters. However, the main purposes of the PDNA are to assess the impact of the disaster and identify priority areas to support the affected government in its short-term response and long-term recovery efforts as well as to recommend interventions necessary for fast recovery (GFDRR 2013). The PDNA encompasses two perspectives: the quantification of physical damage and economic loss and the identification of socioeconomic recovery needs based on information collected from the people in the disaster-affected areas. Thus, the PDNA is a mechanism for joint assessment, medium-term funding strategy, and long-term recovery planning after a disaster.

Many national and international experts and consultants are involved in preparing a single PNDA report; it is generally prepared within one to four months after the disaster. Ideally, a PDNA should commence within the first weeks after the onset of a disaster. Both primary and secondary data are used in this report. The report is generally prepared under the leadership of a relevant ministry of the affected country and draws on all other ministries and departments, as well as international actors in a joint process. For example, the PDNA report for the 2015 Nepal Earthquakes was prepared under the leadership of the National Planning Commission with the active participation of different ministries of the government, private sector, and civil society along with bilateral and international donors and agencies like the ADB, EU, Japan International Cooperation Agency, U.N. Resident and Humanitarian Coordinator, and the World Bank. More than 250 national and international experts were involved in assessing earthquake damage and loss, preparing the two-volume PDNA report within a month of the event. Their findings were presented at the International Conference on Nepal's Reconstruction (GoN 2015).

A PDNA comprises a Damage and Loss Assessment (DaLA), a Human Recovery Needs Assessment (HRNA), and a Recovery Framework. The DaLA is quantitative and is used to estimate damages arising from a disaster, as well as the subsequent economic losses caused by the event. This methodology was first introduced in 1972 by the United Nations Economic Commission for Latin America and the Caribbean (ECLAC). It has since been improved in collaboration with WHO, Pan American Health Organization (PAHO), World Bank, Inter-American Development Bank (IDB), UNESCO and ILO to approximate damage and losses due to disaster events (CRED 2011). The DaLA provides a standardized tool for the valuation of disaster damage and losses.

Although both damage and loss are disaster effects, these two terms have different meanings within the disaster literature. Damage is the value of total and/or partial destruction of physical assets, capital stock, and material goods in the affected area. It occurs during the event itself and immediately after the event; this damage can be repaired or replaced to the same condition before the disaster (Paul 2011). Typically damage includes destruction of houses, crops, roads, bridges, and electrical and water supply systems. Loss, on the other hand, is defined as the negative effects of disaster that cannot be repaired or restored; it is the medium-term effect of disaster. Disaster loss overlaps with disaster damage (Fig. 2.3), and losses lead to changes in economic flows by reducing production, increasing operational costs, and reducing

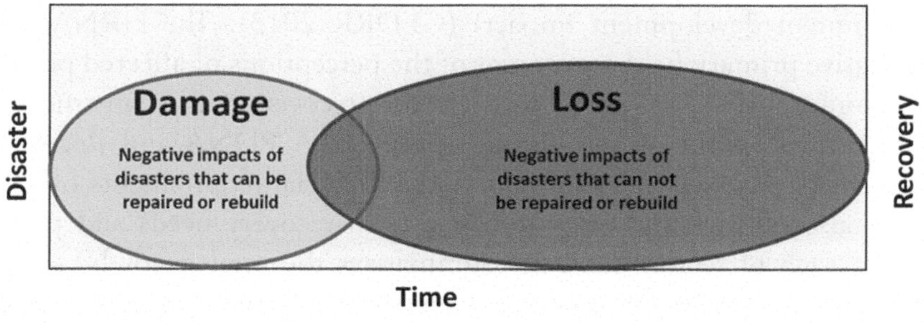

Fig. 2.3 Sequence of disaster effects: Damage and loss (*Source* Compiled by the author)

revenues from providing services (World Bank 2010). Losses continue until full recovery is achieved, which, in some cases, takes several years (CRED 2011). Thus, losses span a longer time than the damage and are valued at current prices. However, the DaLA highlights the possible consequences to the growth of the national economy, and the impact caused by the event in reduced income and livelihoods of individuals and households (GFDRR 2013).

Four economic sectors are considered in the DaLA. The label of economic sectors and subsectors may differ from country to country. The four sectors used for the 2015 Nepal Earthquakes effects were: (1) social sectors (housing and human settlement, health, education, and cultural heritage); (2) productive sectors (agriculture, irrigation, commerce, industry, tourism, and finance); (3) infrastructure sectors (electricity, communications, community infrastructure, transport, and water and sanitation); and (4) crosscutting issues (governance, disaster risk reduction, and environment and forestry) (GoN 2015).[8] In addition to dividing disaster effects into damage and losses, the effect is also disaggregated into private and public sectors (Table 2.4). The share of effects of the 2015 earthquakes in Nepal across sectors is illustrated in Fig. 2.4.

The HRNA focuses on the overall socioeconomic impact of disasters and analyzes the effects of disasters on local patterns of life, social structures, and institutions. The impact analysis is based, among other things, on assessing two important components of the effects of a disaster: economic impact at macro- and micro-levels and the impact of the disaster on the quality of human life in the medium- and long-term (i.e., human development impact) (GFDRR 2013). The HRNA is a qualitative primary field assessment of the perceptions of affected people and communities that provides insight into recovery and reconstruction from the viewpoint of the disaster survivors. A PDNA and Recovery Framework (PDNA/RF) summarizes the financial requirements by sector. It also outlines the early and long-term recovery needs and priorities of each of sector. Table 2.5 summarizes the total needs by sector and subsector for Nepal to recover from the 2015 earthquake. The table shows that the country needs $6.7 billion to recover from the 2015 earthquake.

Table 2.4 2015 Nepal Earthquakes: Summary of disaster effects

Sector	Disaster effects (US$ million)			Distribution of disaster effects (US$ million)	
	Damage	Loss	Total	Private	Public
Social sectors	*3.550*	*536*	*4086*	*3632*	*454*
Housing and human settlements	3036	469	3505	3505	–
Health	64	11	75	14	62
Education	281	33	314	24	290
Cultural heritage	169	23	192	89	102
Production sectors	*581*	*1200*	*1781*	*1581*	*200*
Agriculture	164	119	283	258	26
Irrigation	4	–	4	–	4
Commerce	90	79	169	170	–
Industry	84	109	193	193	–
Tourism	189	624	813	751	61
Finance	50	269	319	209	109
Infrastructure sectors	*524*	*143*	*667*	*173*	*495*
Electricity	178	34	212	156	57
Communications	36	51	87	17	70
Community infrastructure	33	–	33	–	33
Transport	172	49	221	–	221
Water and sanitation	105	9	114	–	114
Crosscutting issues	*519*	*11*	*530*	*18*	*512*
Governance	187	–	187	–	187
Disaster risk reduction	2	–	2	–	2
Environment and forestry	330	11	341	18	323
Total	5174	1890	7065	5404	1661

Source Prepared by the author based on GoN (2015, p. xiii)

However, a PNDA was first published in April 2008 in Bangladesh after a Category IV Tropical Cyclone Sidr hit the country on November 15, 2007 (GoB 2008a). In that year, five PNDA reports were published for five major disasters occurring in Bangladesh, Bolivia, Madagascar, Myanmar, and Yemen; all were published as a Joint Damage, Loss, and Needs Assessment (JDNLA). In the following year, its name changed to PDNA. Since the first report was published, 43 other PNDA reports have been published up to the end of 2016. With nine reports, the highest number of PNDA reports was published in 2009 (GFDRR, n.d.).

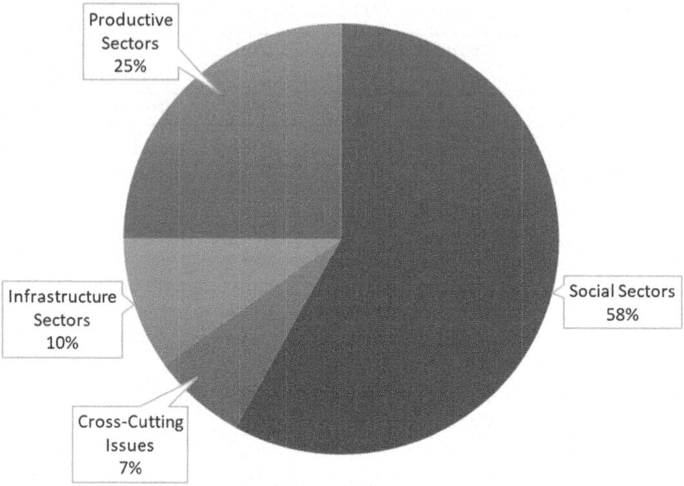

Fig. 2.4 Share of the 2015 Nepal Earthquakes effects across sectors (*Source* Adapted from GoN (2015, p. xiv))

All reports used consistent methodology in assessing damage and losses, but there has been some confusion about the clusters and sectors. Both should use common definitions for selected key terms, such as affected population and vulnerable groups (CRED 2011). Preparing a PDNA requires team members to participate full time from start to finish to prepare a timely PNDA report, which should be based on accurate data. A PNDA is unquestionably useful in strengthening disaster risk management, but a PNDA costs time and money, often exceeding $100 million. Although the cost is generally incurred by international agencies, there is a need to examine whether its preparation is cost-effective given the fact that its preparation coincides with the early response stage of the disaster management cycle when disaster survivors need the urgent attention of all responders.

In addition to the PDNA, affected government and international agencies also publish assessment reports. For example, two months before the PDNA for Cyclone Sidr was published, the Government of Bangladesh published a report, *Super Cyclone Sidr 2007: Impacts and Strategies for Interventions*. The main purpose of this report was "to

Table 2.5 Summary of total needs as suggested by the 2015 Nepal Earthquakes PNDA report, Vol. 1

Sector	Total needs (US$ million) (%)
Social sector	*4077 (60.9)*
Housing	3278 (49.0)
Health	147 (2.2)
Nutrition	50 (0.8)
Education	397 (5.9)
Cultural heritage	206 (3.1)
Productive sectors	*1156 (17.3)*
Agriculture	156 (2.3)
Irrigation	5 (0.1)
Commerce	201 (3.0)
Industry	74 (1.1)
Tourism	387 (5.8)
Finance	335 (5.0)
Infrastructure sectors	*743 (11.1)*
Electricity	186 (2.8)
Communications	49 (0.7)
Community infrastructure	45 (0.7)
Transport	282 (4.2)
Water and sanitation	181 (2.7)
Crosscutting issues	*719 (10.7)*
Governance	184 (2.8)
Disaster risk reduction	82 (1.2)
Environment and forestry	252 (3.8)
Employment and livelihoods	125 (1.9)
Social Protection	64 (1.0)
Gender and Social Inclusion	11 (0.2)
Total	6695

Source Prepared by the author based on GoN (2015, p. xix)

facilitate discussion leading to the finalization of a government endorsement strategy for early recovery and beyond" (GoB 2008b, iii). After the 2015 Nepal Earthquakes, the IASC published a report on mental health and psychological support in May of the same year (IASC 2015). Besides the PDNA, international agencies like WHO also prepare rapid needs assessment for WASH, a subsector of infrastructure sector (see http://globalhandwashing.org/wp-content/uploads/2015/04/WHO-SEARO-Rapid-Needs-Assessment.pdf). An Internet search also indicates that some organizations are referring to the next-generation damage and post-crisis needs assessment tool for reconstruction and recovery planning (see http://reconass.eu/Deliverables/RECONASS_D1.1_State_of_the_Art_of_Assessment_tools_and_preliminary_user_requirements_V1.00.pdf). Therefore, duplication is almost certain, and that duplication of effort should be avoided when both manpower and monetary resources are needed to adequately response to a disaster and prepare for rapid recovery.

Providers of Relief Aid

Humanitarian actors, specifically providers of disaster relief, are numerous and represent a wide array of groups and individuals in terms of geographic location, culture, purposes, interests, mandates, capacity, and logistics expertise (Balcik et al. 2010).[9]

Typology of Providers

Key players in disaster relief can be categorized as follows: governments of affected and non-affected countries, including the domestic and foreign military; multinational and bilateral aid agencies; local, regional, national, and international NGOs; IGOs; individual volunteers both from affected and non-affected countries; domestic and international corporations; and private-sector companies.[10] Often new NGOs and ad hoc committees form just to collect relief supplies and address one particular disaster (Cozzolino 2012). Some humanitarian actors may use their own resources and/or collect donations from public. Because of this diversity, providers of disaster aid are variously classified. For example, Versluis (2014) classified providers as formal and informal. Formal assistance comes from government; multilateral organizations like the World Bank, the African Development Bank (AfDB), the ADB, and the Inter-American Development Bank (IDB); and official aid agencies. Informal assistance comes from expatriates, family, neighbors, friends, and other, smaller local groups like churches.

Like Versluis (2014), Paul (2011) dichotomizes providers of disaster relief into two broad categories: external and domestic. The former includes all entities from unaffected countries that provide emergency assistance to the survivors of disaster-affected areas or host country. Domestic providers comprise all entities of the affected country like national, state, and local governments, domestic NGOs, and private organizations as well as neighbors, friends, and relatives of both affected and unaffected communities. Generally, all members of a disaster-affected community are not impacted equally by disaster. Less affected or unaffected community members may offer relief materials to disaster survivors (Paul et al. 2017). These members are often called "secondary victims" of disasters (Taylor 1989). However, external providers of disaster aid are here known as "international actors" while their domestic counterparts are often called "local actors." Randolph Kent (1987) called the former "international relief network." For any given disaster, local people and organizations are always the first to respond to crisis. They offer immediate assistance (food, shelter, clothing, cash, and labor) and are both providers and distributors of disaster aid. Additionally, they provide crucial psychological support, especially for those who experience death or serious and life-threatening injury in their households (Paul 2011).

Instead of using a dichotomous classification, Alessandra Cozzolino (2012) divided key providers of disaster aid into seven broad entities: (1) governments, (2) the military, (3) aid agencies, (4) donors, (5) NGOs, (6) logistics companies, and (7) other companies. The first entity includes governments of both affected and non-affected countries. These governments have the power to authorize relief operations and mobilize resources to provide aid to disaster survivors of the host country if the host government allows other governments to participate in disaster relief efforts. How much foreign countries should be involved in disaster relief is often dictated by the host country. The engagement of other countries is a delicate matter because other countries can be facilitated or blocked depending on the relationship between the host government and the international community.[11]

Aid agencies and IGOs are actors through which governments of disaster-affected countries can alleviate the suffering of survivors caused by natural as well as man-made disasters. Large agencies are generally global actors (e.g., the World Bank and World Food Program), but many relatively small regional and country-specific aid agencies like the ADB also provide relief. Donors provide most funding for major relief activities either in cash or in kind. Since in-kind donations tend to come from the private sector, public agencies usually fund relief operations financially (Balland and Sobhi 2013). NGOs include several disparate actors, ranging from influential international players like CARE, IFRC, and Oxfam, to small groups and micro-organizations within local communities that can also operate at the international level. Private-sector companies (logistics and others companies) are increasingly present in the humanitarian relief environment.[12] In humanitarian logistics, companies can actively donate relief supplies, collect supplies, and directly provide relief. Foreign NGOs and charities usually collect donations in selected stores and offices, as well as from the general public using selected stores, and/or through phone or online requests (Paul 2011).

Today, almost every country in the world participates in providing humanitarian assistance. This contrasts with the common belief that only rich countries provide disaster relief. Thus, humanitarian action is common among both developed and developing nations (Harmer and Martin 2010). The Financial Tracking System (FTS) of the United Nations OCHA (OCHA 2011) listed more than 180 donors as providers of humanitarian aid in 2010.[13] Other than humanitarian reasons, disaster requires little negotiations and less coordination, unlike long-term development aid (Fuchs and Nils-Hendrik 2012). For this reason, many countries, irrespective of their level of development, participate in offering disaster aid. For example, after the January 2010 Haiti Earthquake, as many as 175 countries provided emergency relief. Eleven of these countries were poorer than Haiti. Within one day of the earthquake's occurrence, 30 countries made aid commitments, including China and low-income countries like Nicaragua (OCHA 2011).

Countries providing emergency aid are classified by Fuchs and Nils-Hendrik (2012) as either "traditional" or "new" donors.[14] The former include developed countries, most of them members of the OECD.[15] "New" donors include developing and socialist countries. Kragelund (2010) proposes the use of the term "non-traditional donor." However, new donors often provide much more emergency aid than traditional donors. For example, after the 2008 floods in Yemen, OECD donors, including Canada and Germany, provided a negligible share of the aid delivery, while new donors like Kuwait, Saudi Arabia, and Singapore contributed more than 90% of the delivered aid amount (Fuchs and Nils-Hendrik 2012). Furthermore, new donors are not only catching up with traditional donors in the amount of their contributions, but are also often among the first to respond to disasters. For example, after committing themselves, China and Canada were the fastest to respond to the devastating 2010 earthquake in Chile. Similarly, Russia and Germany were the first donors to commit to providing emergency relief to Iran after the 2006 earthquake (Fuchs and Nils-Hendrik 2012).

Reasons for Making Donations

Not all humanitarian actors donate disaster relief materials in the form of in-kind and/or cash for the sole purpose of helping others in desperate need or with no expectation that donors will benefit from such donations. Donor intention falls under prosocial behavior (Nelan 2016), implying that all prosocial behaviors are not necessarily selfless. People and organizations often donate, particularly goods, to receive tax breaks (Ariely et al. 2009; Small and Cryder 2016). Instead of throwing goods away, donating those goods that garner tax deductions for donations is rational behavior (Stecklow 2005). In addition to monetary incentives to donate, some donors are set on achieving social status (Small and Cryder 2016). Others are motivated to donate because of media influence. Yet others, both people and organizations, are motivated to donate to clear out closets and storage, to get rid of unwanted household goods like old clothing and clothing of irregular sizes or goods that are close to their expiration dates (Nelan 2016).

Those who donate with the exclusive intention of reducing the sufferings of the survivors of disasters or emergencies show benevolent prosocial behavior, while those who donate to gain personal benefits are showing egoistic prosocial behavior. People belonging to the latter group often want to clear storage spaces by donating goods that are old as well as no longer useful (Nelan 2016). By donating such items, they can feel good about themselves or get satisfaction even if their motive is not truly altruistic (Feigin et al. 2014). Nelan (2016) found materialism was closely associated with donations of goods in response to disasters or for other charitable reasons, particularly for egoistic donors who purge "unwanted goods that may have been purchased with materialistic motivations" (Nelan 2016, 101). By donating excess and inappropriate relief materials, egoistic donors often slow the flow of necessary items to disaster-affected areas. This results in convergence of materials (see Chapter 5).

Two internet sources provided 10 reasons why one should make charitable donations (http://www.800charitycars.org/news/10-reasons-why-you-should-make-charitable-donations/) and why people do not donate to charities (https://www.thelifeyoucansave.org/learn-more/common-objections-to-giving).[16] These reasons are listed in Table 2.6. Although these two sources are referring to donations in a general sense, these same reasons are also applicable in a disaster context.

Table 2.6 Ten reasons why one should make and do not make charitable donations

Number	Should make donation	Should not make donation
1	Paying it forward	Already gave enough, prefer to invest in own community
2	Setting an example	Need to save money for him/herself and for family
3	Spreading love	Personal donation is too small to make a difference
4	Making a world a better place	Poverty cannot be solved
5	Gaining respect and leaving a legacy	Poverty is not the problem—it's the symptom
6	A sense of empowerment	Donations do not help the people who need the most aid
7	Chance to double your donation	Poverty relief is the responsibility of governments
8	Showing gratitude	People should solve their own problems
9	Supporting a cause	Giving aid makes developing countries dependent upon foreign resources and funding
10	Tax deduction	Giving leads to overpopulation

Sources See text (p. 70)

The International Humanitarian System

In the context of providers of humanitarian assistance, some authors, such as Davey et al. (2013), placed providers into a system called the international humanitarian system. This system is defined "as the network of national and international provider agencies, donors and host-government authorities that are functionally connected to each other in the humanitarian endeavor and that share common overreaching goals, norms and principles" (Taylor et al. 2012, 8). This system also includes actors (e.g., military and private-sector entities) for whom humanitarian assistance is not their central mission although they often participate in disaster response. This system is not a formal one because a wide range of agencies and organizations included within system. Thus, the system may be better considered a complex network, constantly adapting and evolving, with different features and characteristics to fit different situations. Some humanitarian responses are led and coordinated by the United Nations and its relevant agencies, while in other cases, the government of a disaster-affected country leads and carries out relief operations with support from donors.[17]

Taylor et al. (2012) categorized all actors of the international humanitarian system as core and non-core actors. The former includes (1) international NGOs (both specifically humanitarian organizations and multi-mandated organizations that work in humanitarian assistance); (2) U.N. agencies, offices, funds, and programs that contribute to humanitarian assistance; (3) the IFRC and its national counterparts; (4) national NGOs (that partner with and/or receive funding from international humanitarian entities for humanitarian operations) as well as local NGOs and local civil society organizations (CSOs); (5) host-government entities (such as interior ministries and National Disaster Management Authorities), that oversee receipt and implementation of international humanitarian assistance; (6) regional, intergovernmental agencies engaged in humanitarian activities; and (7) donor-government agencies/offices that provide financing for international humanitarian operations and institutions.

The non-core humanitarian actors work in parallel to and often coordinate with the rest of the humanitarian system, but they have different goals and approaches. These include militaries, private-sector entities like commercial contractors, religious institutions (differentiated from faith-based operational aid organizations), and diaspora groups as well as formal and informal private donors.

Not only are a large number and variety of actors engaged in disaster relief operations, each has its own missions, interests, capacity, and logistics expertise. Typically, no single actor has sufficient resources to respond effectively and efficiently to a major disaster. However, irrespective of core or non-core status, all of these actors are not providers and distributors of emergency relief materials among disaster survivors. Distributors are discussed next.

Distributors of Disaster Aid

Actors and Disaster Logistics

As noted, not all humanitarian actors directly participate in post-disaster relief operations. Those who do participate are called logistic service providers (LSPs) or logisticians (Nelan 2016). These providers are responsible for the cost-effective flow of relief goods and store and manage donated goods and materials before they distribute donated goods among disaster survivors, with a goal of timely distribution (Thomas and Kopczak 2005). Their functions include warehousing, inventory management, and transporting emergency supplies from entry points like airports and seaports of the disaster-affected country (Coyle et al. 2008). LSPs are tools used to integrate the supply chain and create competitive advantage through increased service and flexibility (Skjoett-Larson 1999; Fabbe-Costes et al. 2008). Planning and execution of the flow of donated relief goods is now increasingly dubbed humanitarian logistics, logistics in humanitarian aid, or (disaster/relief) supply chain. To Ronald Ballou (2007), logistics refers to collecting appropriate emergency materials and delivering them to the right person at the right time (also see Van Wassenhove 2006). U.N. DHA (1992) defines logistics as "The range of operational activities concerned with supply, handling, transportation and distribution of materials" (cited in ReliefWeb 2008, 34). Several authors (e.g., Trunick 2005; Van Wassenhove 2006) claim that more than 80% of actions humanitarian organizations take in disaster relief operations are related to logistics.

In recent years, logistics in disaster aid has become more important topic because humanitarian actors want to make supply chains successful. Currently, logistics confront many challenges because excessive and inappropriate relief materials arrive in disaster-affected areas along with high-priority materials. The unrestricted flow of different types of donations can create, "a huge quantity of items, an extremely heterogeneous flow, arriving within a short time span to an area with limited space, resources, and personnel to process and distribute them to their intended recipients, and people in great need" (Holguin-Veras et al. 2014, 2). The influx of unnecessary and non-priority emergency materials, called the convergence phenomenon in disaster literature, is further discussed in Chapter 5. Such an influx can cause a supply chain to become congested, resulting in delays in necessary items reaching disaster survivors (Nelan 2016).

Ideally, LSPs manage and facilitate the supply chain (Chandes and Pache 2010). LSPs must be transparent, and their actions must not be interpreted as favoring one set of individuals, households, or communities at the expense of others. They also need to understand the donation process, which starts when donations enter the chain and ends when disaster survivors receive the donations. The duration from beginning to end is crucial because survivors needs change over time. That change should be reflected in donations. Specific needs and the donations responding to those needs are not stable (Nelan 2016). If the supply chain takes too long to deliver donations, some materials needed at the early stage of disaster response are no longer needed. Needs change as the relief effort evolves. Needs also vary by location. Time and urgency are critical in distributing relief goods, particularly during the early stage of disaster response. At this stage, the logistics should provide a continuous flow of appropriate and urgently needed relief goods in a short time.

LSPs generally comprise a large number of individuals and organizations. As a result, to make the supply chain successful, LSPs must overcome many problems to maintain the flow of donated goods. Delivering some emergency supplies, particularly during the early disaster response phase, is invariably delayed in major disasters for several reasons. Supply chain at this stage may be interrupted because of damage to infrastructure in the disaster-affected areas, causing significant challenges in transporting critical supplies. Many major disasters cause infrastructure damage that cause delays, including Hurricane Mitch in 1998, Hurricane Katrina in 2004, IOT in 2004, the Haiti Earthquake in 2010, the Nepal Earthquake in 2015, and the 2017 Hurricane Irma in Puerto Rica. Shortages of drivers and trucks, not to mention the imposition of curfews in disaster-affected areas, are also reasons for delays in delivering donated supplies (Wachtendorf et al. 2013; Paul et al. 2017).

Moreover, lack of coordination and partnership among the many and various actors involved in disaster relief operations is a key reason for failure to alleviate the sufferings of disaster survivors. Lack of coordination is caused by the sheer number and variety of LSPs, scarcity of resources, and lack of experience and training among some or many LSPs (Balcik et al. 2010). Major disasters, after all, do not occur often and are noted for their lack of predictability, so many of those who are part of the humanitarian supply chain may have little experience. Those who helped in previous disasters may have retired, moved away, or are physically unable to participate in the chain (Nelan 2016), leaving less experienced people responding to a current disaster. As indicated, moreover, LSPs have different backgrounds, cultures, purposes, and mandates. They must work together to successfully complete the relief operation. Their coordination is needed to deliver aid in a timely manner, reduce duplication of emergency goods, and increase delivery effectiveness and efficiency, a task difficult enough without such variety.

Ways to Improve Humanitarian Logistics

Many researchers have suggested ways to make disaster assistance supply chains more effective and efficient. Holguin-Veras and Jaller (2012) proposed establishing pre-positioned areas where donations can be stored in preparation for future disasters. Experienced agencies often have a shortlist of hotspot regions (such as central Africa for conflicts or Middle

America and the Caribbean for hurricanes). They then pre-position supplies at major airfields so they can respond more rapidly. This strategy works better for rapid-onset disasters like earthquakes and tornadoes (Van Wassenhove 2006; Duran et al. 2013). Slow-onset disasters like hurricanes and droughts allow enough lead time for organizations and LSPs to move and store emergency supplies to areas that will be affected.

Nelan (2016) reported that before Hurricane Sandy, one state agency in New Jersey opened up a donations warehouse that would be able to respond quickly after the storm. She further reported that a large NGO in Oklahoma began to stocking a warehouse in preparation for tornado season. By pre-positioning supplies, they anticipated needs and could immediately meet them. However, pre-positioning supplies in advance of disasters are not always cost-effective because it requires substantial maintenance costs and risks wasting medical and food supplies as they expire. For this reason, Holguin-Veras and Jaller (2012) maintain that all disaster survivors will not need every item, and therefore they suggest stocking the highest priority supplies.

Ito et al. (2014) recommended five ways to achieve successful logistics during a disaster relief operation in Japan. These recommendations can also apply to other countries. First, they recommended having separate storage spaces for emergency goods and for other goods not immediately necessary. This separation can avoid confusion and thus reduces delay. In Japan, choosing relief items from a warehouse requires a relatively long time because warehouses store both emergency and non-emergency goods. Relief goods are generally taken in the country from primary stockyards, delivered to secondary stockyards, and only later are sent to shelters where disaster survivors can be received the goods (Fig. 2.5). Primary stockyards are located at prefectural levels, the secondary stockyards are at municipality levels, and the shelters are at the ward levels. Problems with logistics arise if any donor sends relief goods directly to secondary stockyards and/or shelters, thus bypassing the primary stockyards (Ito et al. 2014). Ito et al. (2014) suggested that the government of a disaster-affected country and international actors should establish a standard system for storing relief goods. Second, they recommended introducing a system to control the quantities of incoming goods from foreign countries.

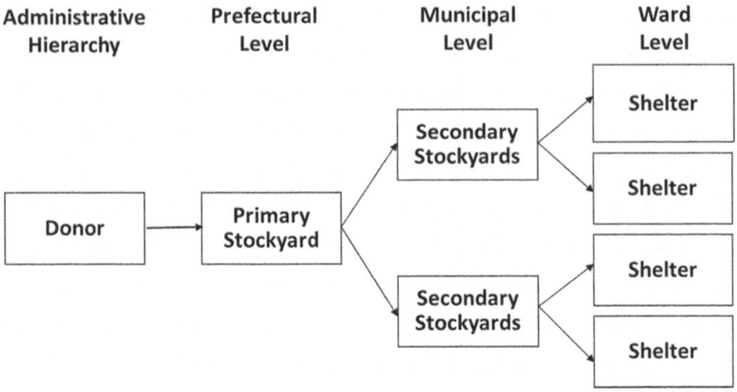

Fig. 2.5 Flow of relief goods in Japan (*Source* Modified after Ito et al. (2014, p. 144))

The third recommendation is to control the quality and types of incoming goods from donors to keep unnecessary goods from disrupting the relief operation. Ito and his colleagues (2014) also recommended reducing time required for customs to process goods coming from foreign countries. Customs delays often mean some goods from foreign countries reach disaster survivors too late, which happened in the immediate aftermath of 2015 earthquakes in Nepal (Paul et al. 2017).[18] The fourth recommendation is to improve efficiency at the primary and secondary stockyards by (1) standardizing the packaging for ease of handling, (2) standardizing the codes on the packages, (3) using a unit load system to reduce operation time, and (4) sharing resources with the disaster-affected country, which may not have sufficient equipment, vehicles, or skilled staff. Donor countries should be prepared to supply such resources. The final recommendation is to increase work efficiency using information technology.

Large organizations like the American Red Cross and the United States Agency for International Development (USAID) recommend donations of cash over goods to stop the flow of unsolicited relief materials and reduce the volume of such donations. Nelan (2016) claims that material donations can undermine local suppliers and can also bring in emergency items that are not useful within the culture of the disaster-affected communities or climate of the affected areas. Moreover, apart from safety issues, food donations can be problematic because different religions have different dietary restrictions. Similarly, clothing may not be useful because of cultural restrictions, tradition, or climate (Harvey 2007; Kelman and Dodds 2009).[19]

Ozeguven and Ozbay (2013) proposed controlling incoming goods by using modern technologies like Radio Frequency Identification Devises (RFIDs) to track donated goods as they are transported to disaster-affected areas.[20] This could slow or stop low-priority materials from entering into the area affected by a disaster. Aidmatrix can be used to communicate directly and quickly the needs of disaster survivors to humanitarian actors. Aidmatrix is an online network developed in 2000 by a supply chain software company the Aidmatrix Foundation, Inc. Aidmatrix partners with governments, donors and relief agencies worldwide to help those organizations respond quickly, with quality aid that matches local needs (Aidmatrix 2016). This online network can make the supply chain more agile, which is defined in humanitarian logistics as "rapid deployment on demand" (Van Wassenhove 2006, 486).

Changes in survivors needs and lack of predictability of type, quantity, quality, source, and arrival time of donated goods to disaster-affected areas is a serious problem in the logistics of a humanitarian supply chain. The unpredictability can be overcome by adjusting the design of the supply chain, mostly making the supply chain flexible and dynamic (Lee 2004). Adaptability is necessary to help adjust the flow of donations in uncertain and complex situations (Nelan 2016). Adaptability often requires people and organizations to stop inappropriate or excessive donations before they enter affected areas. Although not accepting donations from donors is difficult, LSPs must exclude non-priority materials, focusing on necessities. However, LSPs do not want to displease donors because they want to keep donors in the supply chain for future disasters (Apte 2010; Nelan 2016).

A report published by Friz Institute (Thomas and Kopczak 2005) recommended five strategies to improve to improve humanitarian logistics: (1) creating a professional logistics community to share knowledge and experience on common issues and to serve as a consistent, powerful voice with all stakeholders in the sector; (2) investing in standardized training and certification to build a pool of logistics professionals;

(3) focusing on metrics and performance measurement to empower logistics expert to demonstrate and improve the effectiveness of humanitarian supply chains; (4) communicating the strategic importance of logistics so logistics expert can highlight the contributions of logistics, thus making it possible to obtain funding and resources; and (5) developing flexible technology solutions to improve responsiveness and increase the effectiveness of people and processes. Although these recommendations were made in 2005, little progress has been made in implementing these recommendations.

Domestic and Foreign Military Forces

Depending on the magnitude of a natural disaster, both domestic and foreign military forces often provide resources and military assets during humanitarian operations: fixed-wing aircraft and helicopters, skills, assistance in installing mobile hospitals, preventing disease outbreak, supplying pure drinking water and other essential supplies, establishing telecommunications, and repairing roads and bridges. Military help is most frequently needed for airlift operations to transport personnel and relief supplies to disaster sites. The second and third most frequently deployed assets are medical support and expert personnel (Wiharta et al. 2008). In essence, military forces are more distributors of disaster relief than providers. In some countries, such as the UK, it is normal for the army to participate in responses to natural or man-made disaster. In many countries, however, people seriously object to such participation, particularly among the elites of a society (Wiharta et al. 2008).

Although the USA is the most proactive in using military assets for disaster response, after Hurricane Katrina, the issue was vigorously debated among Americans.[21] Many felt that the main purpose of the military is to defend the country from external threats and safeguard vital national interests, not to participate in disaster relief efforts and development-related tasks. This notion is widely prevalent in other countries among public and humanitarian actors alike. Despite concern with trespassing on humanitarian space, however, the humanitarian community has begun to accept more and more that military assets can be a vital part of the response to natural disasters. Using both domestic and foreign military assets has advantages, particularly in developing countries where disaster assistance does not always reach the people who desperately need it (Paul 2011). Domestic military also shares these advantages, so here both domestic and foreign military are presented in the discussion.

Foreign military forces require permission from the host country to directly participate in disaster relief operations. China and India, however, do not have stated policies about accepting foreign military assets, and certain other countries have policies limiting the use of military assets in international disaster response. Since 2000, foreign military was most engaged in providing humanitarian assistance during the disaster responses to the 2004 IOT in Aceh, Indonesia, and the 2005 South Asian Earthquake in Pakistan (Wiharta et al. 2008). The USA has deployed its military assets most frequently and in the greatest numbers since 2000 (see Wiharta et al. 2008). The issue of having foreign militaries participate in international disaster relief assistance is under hot debate because such assistance is traditionally considered the domain of civilian humanitarian agencies. Moreover, disaster-affected countries may consider accepting such assistance a breach of national security.

Deployment of foreign military in disaster response is inevitably a political decision, both by the contributing country and by the host country. The host country has the right to reject any offer of military assistance during a disaster, even if the contributing country offers as an act of friendship or solidarity. Several non-binding guidelines and essential principles regulate direct or indirect participation of foreign military forces in disaster response. The contributing country must request the disaster-affected country allow such help, gaining the consent of the affected country. The mission should be under the overall control of a civilian authority and of limited duration; moreover, the disaster-affected country should not bear any expense incurred by the foreign military force. The foreign force should be unarmed, but in national uniforms, it must respect the sovereignty of the affected country and it must acknowledge its primary responsibility is to provide humanitarian assistance. Both the receiving and contributing countries should follow the Oslo Guidelines, which were formulated in 1994 and updated in 2006 (OCHA 2007). The updated report contains 107 guidelines.

The Oslo Guidelines stipulate that all humanitarian assistance must be provided in accordance with the U.N. core principles of humanity, neutrality, and impartiality (paragraph 20). Although the guidelines are designed to govern foreign military forces participating in relief operations only in peacetime, military forces have also participated in several major natural disasters that occurred in conflict areas, such as Aceh Province in Indonesia and Sri Lanka in the case of the 2004 IOT, the 2005 earthquake in Kashmir in Pakistan, and 2013 Typhoon Haiyan in the Philippines. The Oslo Guidelines recommend using domestic military forces as a first resort only if the country lacks a civilian alternative and only if using military assets can meet a critical humanitarian need. The disaster-affected country should use foreign military forces only as a last resort. These guidelines also state that foreign forces and civil defense should complement each other to provide specific needs (paragraph 24). The principles of Good Humanitarian Donorship affirm the primary position of civilian organizations in implementing humanitarian action, and the disaster-affected country must ensure that foreign military assistance is in conformity with international humanitarian law and humanitarian principles and recognizes the leading role of humanitarian organizations (Hofmann and Hudson 2009).

Using domestic and foreign military forces for relief operations does have many advantages, particularly in developing countries. First, they have a variety of different capabilities: air and sealift capacity, logistics management, engineering support, communications, and medical assistance capabilities. They are fair and quick in disbursing disaster aid (Paul 2011). They are well equipped and sufficiently trained to effectively handle post-disaster situation, which can save lives. These forces can repair roads, build houses and bridges, and carry relief goods to the disaster site very quickly.

The presence of foreign military forces is, however, generally perceived by the residents of disaster-affected countries as a serious threat to national sovereignty. These forces often face difficulty in maintaining neutrality with no provision for monitoring their activities. Additionally, deployment of foreign military forces is often delayed because they must obtain political clearance to enter the disaster-affected country, and they must transport heavy equipment by sea and land, which takes time. This slows the start of humanitarian operation. Although the Oslo Guidelines provide clear guidance, command is a recurrent problem: under whose command will foreign forces work in disaster-impacted areas? The lack of coordination between domestic and/or foreign military personnel and other humanitarian actors, including local people, hinders smooth relief operation. For this reason, the Red Cross and Red Crescent Movement has its own guidelines for relations with the military. The key principles are to: (1) maintain a dialogue with armed forces; (2) ensure that the military acts in accordance with the fundamental principles of independence, neutrality, and impartiality; (3) distinguish the respective roles of military bodies and humanitarian actors, paying particular attention to local perceptions as well as the wider public perception; (4) use military assets as a last resort, which can only be justified by serious and urgent humanitarian needs, as well as by the lack of alternatives; and (5) avoid using armed protection (Hofmann and Hudson 2009).

Many humanitarian actors realize that the military force can be vital in supporting humanitarian relief efforts. Given the growing involvement of military actors in relief activities, humanitarian actors have the opportunity and a responsibility to engage more strategically with the military to limit the risks inherent in military involvement and maximize the potential benefits to disaster survivors. The question for humanitarian organizations is no longer whether to engage with the military, but rather how to work with them to facilitate disaster relief, save lives, and relieve the sufferings of survivors.

Coordination, Cooperation, and Partnership

Coordination, collaboration, and partnerships are frequently used interchangeably in the disaster supply chain literature (e.g., Kent 1987; Russell 2005; Balcik et al. 2010). Although each of these terms may be interpreted differently by different researchers, coordination or partnership has continued to be a fundamental weakness in disaster relief operations and in disaster management, particularly in developing countries (Rey 2001; Auzzir et al. 2014). Arguably, the Cluster Approach may not achieve its goals of improving timeliness, effectiveness, and coordination without partnership (Actionaid 2007).Thus, hazard researchers and organizations involved in humanitarian assistance have pursued this topic primarily because of "the increasing scarcity of global resources, accountability concerns, and the potential opportunities provided by advances in global information technologies" (Lindenberg and Bryant 2001, 159). Because logistics accounts for 80% of relief operations (Van Wassenhove 2006), relief chain coordination or partnerships between government and private sectors are crucial to improving relief chain performance as well as making relief operations successful. This is particularly true when governmental oversight and resources are severely lacking (Simo and Bies 2007). This partnership, widely called public–private partnerships (PPP), is a strategy used by government to encourage private participation in public projects (Johannessen et al. 2013). Without partnerships, methods of collecting and distributing goods to those in need can be inefficient and ineffective (Whybark 2007).

Coordination or partnership simply refers to the relationships and interactions among all stakeholders (e.g., government, local communities, NGOs, private sector, as well as donor countries) taking part in relief operations (Balcik et al. 2010). The Global Humanitarian Platform (GHP), formed in 2007, adopted the Principles of Partnership, which should form the foundation for how humanitarian actors work together.[22] The GHP is an international forum created to bring together on an equal footing the three main families of the wider humanitarian community (NGOs, the Red Cross and Red Crescent Movement, and the United Nations) to make humanitarian action more effective (ReliefWeb 2008). Principles of Partnership, as endorsed by the Global Humanitarian Platform in 2007, are listed in Table 2.7.

According to Balcik et al. (2010), the two types of coordination/partnership are vertical and horizontal. Vertical coordination refers to the extent to which an organization coordinates with other organizations above or below its level, while horizontal coordination means the level of cooperation with organizations belonged to the same hierarchy within the chain. Coordination can also be classified by describing the

Table 2.7 Principles of partnership, endorsed by the Global Humanitarian Platform 2007

Principle	Explanation
Equality	Irrespective of their size and power, partnership in the Global Humanitarian Platform must be based on mutual respect. The partners should respect each other's mandates, and recognize each other's independence and commitments
Transparency	All partners should be financially transparent, and they must share relevant information. These will help to increase the trust among all participating organizations
Result-oriented approach	All humanitarian actions should be based on reality. These actions should also be action-orientated
Responsibility	Using their own resources, all partner organizations must perform their assigned tasks in a timely manner
Complementarity	All partners should complement each other's contributions in emergency response. They should also be overcome local language and cultural barriers

Source Prepared by the author based on Actionaid (2007, p. 5)

time-based engagement level as long-term (strategic), medium-term (tactical), and day-to-day (operational).

Using the strength of practical/physical actions or ties, Coles et al. (2016) classified four types of partnerships: (1) hierarchical (low level of action/degree of belief, which implies that the partnership lacks communication and commitment); (2) bilateral (low level of action, but a high degree of belief, meaning the partnership is a good match in terms of perspective but not necessarily a good match logistically); (3) discrete (high level of action and a low degree of belief, reflecting very effective partnership at achieving action-oriented goals although the partnership may not last for long); and (4) recurrent (high level of action and a high degree of belief, the most stable type of partnerships).

Partnership does not necessarily imply centralized decision making (Kent 1987), but partnership is necessary for several reasons. First, partnership can help communicate needs of disaster survivors more effectively and bring donations as quickly as possible to the disaster-affected areas or communities. Partnership is also required to share information, knowledge, experience, and technological resources so that partner organizations can increase productivity and efficiency in relief efforts (Lee et al. 1997; Altay and Pal 2014). Third, partnership or coordination eliminates excessive and inappropriate donations in the supply chain. It maximizes efficiency and reduces delivery times, making disaster relief efforts successful. Finally, it also means conducting joint projects, regional division of tasks, or a cluster-based system in which each cluster represents a different sector area.

Disaster relief operations last for few weeks to a few months, so conflict or distrust may arise among partners during this time. To avoid this situation, partnerships need a common language and understanding of operational approach, resource allocation, and an agency timeline (Coles et al. 2012). Wishart (2008) maintained that the best partnerships come out of relationships where those involved treated all others as equal, not as subordinate. Mitchell (2006) stated that partnerships last longer if they are based on expansive and compelling ideas. Because new organizations often form in the wake of a major disaster, they could add resources or a new perspective to a relief operation. However, these new organizations also vary greatly in reliability, efficiency, and durability. Established organizations need to understand the characteristics of emergent organizations as their partners.

Organizations are often reluctant, however, to pass on information to others because they often compete for donations from the same donors or for sole attention from the media, particularly during the early stages of the relief response (Stephenson and Schnitzer 2006; Bharosa et al. 2010; Wakolbinger and Toyasaki 2011; Nelan 2016). Organization may also hesitate to share information with other organizations if they believe this information gives them a competitive advantage in attracting media and donor attention (Kent 1987; Stephenson 2005). Day et al. (2009) reported that following Hurricane Katrina organizations were unwilling to share data, thereby impeding information flow in the relief effort. Additionally, occasionally organizations simply do not have sufficiently accurate information to share with partners (Kent 1987).

However, Nelan (2016, 20) maintains that "coordination, whether it is through communication or through partnerships and transportation, within the human supply chain could minimize costs, both monetarily and in the challenges caused by excessive and inappropriate goods." Donors often demand more accountability and become less tolerant of inefficiencies in relief, which should strongly encourage relief organizations to collaborate (Thomas and Kopczak 2005). Despite the advantages and donor pressure, organizations often have many problems in establishing partnerships and coordination during disaster relief operations. One important barrier is associated with language and cultural differences among the partner organizations. If they come from different countries and cultures, these differences can impede coordination (Moore et al. 2013).

Because of the many sources of unpredictability in disaster relief, coordination may suffer. First, some rapid-onset disasters, particularly for earthquakes, are unpredictable because it is hard to identify the exact location, timing, and intensity of disaster, specially before a disaster occurs. For slow-onset disasters like hurricanes, potential landfall area and other pertinent information can be predicted. Second, population characteristics and preexisting regional infrastructure in many disaster-prone areas are not be readily available, and so the extent of post-disaster infrastructure damage may not be predictable. The political environment of the disaster-affected country and post-disaster funding levels are also unpredictable. Added to these factors are resource constraints among potential partners and costs associated with coordination; both hinder coordination and partnership before disasters occur.

Social Networks as Providers and Distributors of Disaster Relief

Social networks, a crucial embodiment of social capital, have proved helpful for individuals and households coping with natural disasters in both the response and recovery phases of the disaster management cycle (Murphy 2007; Islam and Walkerden 2014). Members of these networks provide social, financial, and emotional support to disaster survivors in the immediate aftermath of such an event. They also help with response efforts by arranging temporary shelter for disaster survivors, providing access to various resources and information, and distributing food, drinking water, medicine, clothing, building materials, and other necessary goods. These efforts save lives, reduce suffering, enhance the disaster recovery process, and restore lives, as much as possible, to pre-disaster levels (Mileti 1999; Montz et al. 2017). This humanitarian assistance facilitates the transition of survivors from immediate relief to long-term rehabilitation and development (Smith and Ward 1998).

Social networks are relationships that develop between individuals/households and/or groups (e.g., local NGOs and government organizations—GOs) through family ties, friendship, similar interests and beliefs, organizational life, or other types of social connection (Carpenter 2013). Note that individual and household social networks are commonly differentiated into bonding, bridging, and linking networks (Szreter and Woolcock 2004; Aldrich 2011). Each network identifies a different type and strength in relationships and composition of networks and thus different outcomes for individuals, households, and communities. *Bonding social networks* describe the connections among individuals who are emotionally close like family members and close relatives. This strong connection makes this type of social capital an excellent source of social support and personal assistance, especially in times of need (Aldrich and Meyer 2014). In contrast, *bridging networks* includes loosely connected acquaintances or individuals that span social groups, such as neighbors and friends. Members of both networks often provide financial (e.g., loans and gifts for property repair, or purchase of food, clothing, and utensils) and nonfinancial resources (e.g., search and rescue, debris removal, and child care) to disaster survivors. Bonding and bridging social capital work in complementary but distinct ways during and after disasters, and communities regularly have more of one type than the other (Elliot et al. 2010).

Linking networks describe individual or household connections with organizations, including local, regional, and domestic NGOs and GOs at different levels, that have influence over the circumstances of their members (Woolcock 2001). The first two networks represent horizontal relationships, and the last represents vertical relationships (Patulny and Svendsen 2007). Memberships in such organizations improve the chances of quickly receiving necessary disaster relief materials because these organizations often provide humanitarian assistance solely to individuals within their jurisdiction (Aldrich 2011; Paul 2006). Additionally, members of these organizations are more likely to articulate their needs to authorities of such organizations and work together to overcome obstacles in their efforts to return life to pre-disaster level (Yasui 2007; Lin 2008). In contrast, individuals and households who are not members of any NGO are less likely to receive assistance from such organizations (Dynes 2005, 2006). Many hazard researchers claim that disasters can motivate survivors to participate in civil society organizations as well as create new organizations to help neighbors and community members suffering from post-disaster hardship.

This chapter clearly suggests that more humanitarian actors are now engaged in providing disaster relief than in the past, and their number will increase significantly in the future. Humanitarian actors seem competitive about aid, and this competition will intensify in the future. Although many guidelines and principles already exist to permit coordination among these actors, all major disaster relief operations in the past have suffered from coordinate lack of coordination among all actors, leading to duplication of effort, delayed flow along the relief chain, and convergence of unwanted relief materials. Better coordination is necessary at the local, national, regional, and international levels to make disaster relief operations as flawless as possible.

Notes

1. The CAP begins at the country level, led by the RC/HC, which provides analysis of needs and identifies strategic objectives. Cluster Approach is central to the CAP process, which is discussed later in this chapter. CAPs are typically annual appeals, usually launched each November although new CAPS can be issued as needed and old ones can be revised at midyear (OCHA 2012).
2. In their proposals to the Flash Appeal, international organizations listed 83 local NGOs as partners for implementing their plans, meaning Nepalese NGOs were relegated to the role of subcontractor (Troutman 2015).

3. Even if U.N. launches a major international appeal for an emergency, member states may well ignore it.
4. At least 16 United Nations agencies, in one way or another, can become involved in disaster relief activities.
5. IGOs often suffer from lack of clarity, overlapping responsibilities, inter-agency rivalry, and lack of coordination.
6. An earmark is a provision by which a donor designates the specific purpose of a particular donation. For example, a donor or an international relief organization may earmark its cash donation exclusively for buying tents to provide temporary shelter for disaster survivors. The donor expects the IGO channel to comply with this restriction and to report on how the contribution has been used (Kent 1987; Toyasaki and Wakolbinger 2014). Earmarking is essentially a way of organizing and allocating money in the corporate world.
7. The 2005 *Humanitarian Response Review* paper, commissioned by the U.N. Emergency Relief Coordinator (ERC) and the Under-Secretary-General for Humanitarian Affairs, initially identified the Cluster Approach as a way to improve accountability, predictability, and leadership by identifying organizational leads in areas where there was a gap in humanitarian response (Actionaid 2007).
8. Subsectors are listed within parentheses.
9. Strictly speaking, humanitarian actors encompass both actors who donate necessary goods and/or cash to disaster survivors or sufferers of complex emergencies. Generally, the number of actors for the first is higher than for the second, primarily because unlike disasters, complex emergencies may last years or even decades.
10. Based on their role, Cozzolino (2012) classified private-sector companies into three types: donors, collectors, and providers. As a donor, a company can support disaster relief efforts with cash. As a collector, a company can ask for cash donations from customers, employees, and suppliers. As a provider, a company can offer its goods and services for free (in-kind donation) or as part of sales. In disaster relief efforts, when a company is exclusively donor and/or collector, it is simply placed in the donor category. The model refers to the company category only as a provider or acts as a provider in addition to being donor/collector.

11. In many cases, host countries do not have good relations with their neighbors (Cozzolino 2012).
12. Logistic companies manage the flow of items between the point of origin and the point of consumption to meet the requirements of customers. The resources so managed can include food, materials, animals, equipment, and liquids; they can also include abstract items like time and information. The logistics of physical items usually involves integrating information flow, material handling, production, packaging, inventory, transportation, warehousing, and often security.
13. This is a Web-based, searchable tracking system for contributions that reflects all humanitarian funding reported to OCHA, including contributions to CAP, responses to natural disasters, and all other humanitarian aid as reported to OCHA. In-kind contributions, with a dollar value reported by the donor or recipient entity, are also recorded (ReliefWeb 2008).
14. So-called new donors have a long tradition of providing emergency aid to other developing countries.
15. Currently, 35 developed countries are members of OECD, which was formed on December 14, 1960.
16. Last accessed on October 28, 2017.
17. For a history of the international humanitarian system, see Davey et al. (2013).
18. Beginning on May 26, the Nepalese government stopped allowing humanitarian goods to come in duty-free. Other than such items as medicine and tarps, the government imposed import taxes on almost all other incoming humanitarian aid. This action did delay relief efforts, but it also stopped unnecessary and useless relief items from entering the country. However, the Nepalese public whole heartedly supported the imposition of import taxes and review of emergency items coming from abroad (Paul et al. 2017).
19. The advantages and disadvantages of cash donations are discussed in Chapter 5.
20. RFID refers to a wireless system comprising two components: tags and readers. The reader is a device that has one or more antennae that emit radio waves and receive signals from the RFID tag. Tags can be classified as passive or active. Passive RFID tags receive energy from nearby readers and active RFID tags have a local power source.

21. The USA deployed its military assets 15 times in response to overseas natural disasters between 2003 and 2006. Most of these were in the Caribbean and Central America (Wiharta et al. 2008).
22. The Global Humanitarian Platform seeks to (1) acknowledge diversity as an asset of the humanitarian community and recognize the interdependence of humanitarian organizations; (2) build and nurture effective partnership; (3) enhance the effectiveness of humanitarian actions, recognizing that no single actor can cover all the needs, and that collaboration is necessary; and (4) maximize complementarity, because various actors have different mandates, expertise, approaches, and resources (Actionaid 2007).

References

Actionaid. 2007. *The Cluster Approach in Mozambique 2007: Is It Improving Emergency Response? An NGO Perspective*. Johannesburg, South Africa.

Aidmatrix. 2016. The Aidmatrix Network. http://www.aidmatrixnetwork.org/maps/. Last Accessed 7 July 2017.

Aldrich, D.P. 2011. The Power of People: Social Capital's Role in Recovery from the 1995 Kobe Earthquakes. *Natural Hazards* 56: 595–611.

Aldrich, D.P., and M.A. Meyer. 2014. Social Capital and Community Resilience. *American Behavioral Scientists* 59 (2): 220–237.

Alexander, D. 2000. *Confronting Catastrophe*. Oxford: Oxford University Press.

Altay, N., and R. Pal. 2014. Information Diffusion Among Agents: Implications for Humanitarian Operations. *Production and Operations Management* 23 (6): 1015–1027.

Apte, A. 2010. *Humanitarian Logistics: A New Field of Research and Action.* Hanover, MA: Publishers Inc.

Ariely, D., A. Bracha, and S. Meier. 2009. Doing Good or Doing Well? Image Motivation and Monetary Incentives in Behaving Prosocially. *American Economic Review* 99 (1): 544–555.

Auzzir, Z.A., R.P. Haigh, and D. Amaratunga. 2014. Public-Private Partnerships (PPP) in Disaster Management in Developing Countries: A Conceptual Framework. *Procedia Economics and Finance* 18: 807–814.

Balcik, B., B.M. Beamon, C.C. Krejci, K.M. Muramatsu, and M. Ramirez. 2010. Coordination in Humanitarian Relief Chains: Practices, Challenges and Opportunities. *International Journal of Production Economics* 126 (1): 22–34.

Balland, J., and N.A. Sobhi. 2013. *Humanitarian Relief Organizations and Its Relationship with Logistics Service Providers*. Jonkoping, Sweden: Jonkoping University.

Ballou, R. 2007. The Evolution and Future of Logistics and Supply Chain Management. *European Business Review* 19 (4): 332–348.

Batha, F. 2005. How Does Aid Flow After a Disaster? http://alertnet.org/thefacts/reliefresources114244943761.htm. Accessed 4 Aug 2017.

Becerra, O., E.A. Cavallo, and I. Noy. 2014. Foreign Aid in the Aftermath of Large Disasters. *Review of Development Economics* 18 (3): 445–460.

Bharossa, N., J. Lee, and M. Janssen. 2010. Challenges and Obstacles to Sharing and Coordinating Information During Multi-agency Disaster Response: Propositions from Field Exercises. *Production and Operations Management* 23 (9): 1511–1521.

Bradshaw, S. 2013. *Gender, Development and Disasters*. Cheltenham, UK: Edward Elgar.

Carpenter, A. 2013. *Social Ties, Space, and Resilience: Literature Review of Community Resilience to Disasters and Constituent Social and Built Environment Factors*. Atlanta: Federal Reserve Bank of Atlanta.

Chandes, J., and G. Pache. 2010. Investigating Humanitarian Logistics Issues: From Operations Management to Strategic Action. *Journal of Manufacturing Technology Management* 21: 320–340.

Chou, C.H., F.M. Zahedi, and H. Zhao. 2014. Ontology-Based Evaluation of Natural Disaster Management Websites: A Multistakeholder Perspective. *MIS Quarterly* 38: 997–1016.

Coles, J.B., J. Zhuang, and J. Vates. 2012. Case Study in Disaster Relief: A Descriptive Analysis of Agency Partnerships in the Aftermath of the January 12th, 2010 Haitian Earthquake. *Socio-Economic Planning Sciences* 46: 67–77.

Coles, J., J. Zhang, and J. Zhuang. 2016. Partnership Behavior in Disaster Relief Operations: A Case Study of Comparison of the Responses to the Tornado in Joplin, Missouri and Hurricane Sandy Along the Jersey Coast. *Natural Hazards* 84: 625–647.

Collinson, S., M. Buchanan-Smith, and S. Elhawary. 2009. *Good Humanitarian Donsorship Principles in Practice: Assessing Humanitarian Assistance to Internally Displaced Persons in Sudan and Sri Lanka*. London: USAID.

Coppola, D.P. 2007. *Introduction to International Disaster Management*. New York: Elsevier.

Coyle, J.J., C.J. Langley, B. Gibson, R.A. Novak, and E.J. Bardi. 2008. *Supply Chain Management: A Logistics Perspective*. Boston: Cengage Learning.

Cozzolino, A. 2012. *Humanitarian Logistics: Cross-Sector Cooperation in Disaster Relief*. New York: Springer.

CRED (Centre for research on the Epidemiology of Disasters). 2011. *Disaster Loss Characterization: Review of Human and Economic Impact Indicator Definitions*. Brussels: Catholic University of Louvain.

Davey, E., J. Borton, and M. Foley. 2013. A History of the Humanitarian System: Western Origin and Foundation. Humanitarian Policy Group (HPG) Working Paper, Overseas Development Institute, London.

Day, J.M., I. Junglas, and L. Silva. 2009. Information Flow Impediments in Disaster Relief Supply Chains. *Journal of the Association of Information Systems* 10 (8): 637–660.

Duran, S., O. Ergun, P. Keskinocak, and J.L. Swann. 2013. Humanitarian Logistics: Advanced Purchasing and Pre-positioning of Relief Items. *Handbook of Global Logistics: International Series in Operations Research and Management Science* 181: 447–462.

Dynes, R. 2005. *Community Social Capital as the Primary Basis for Resilience*. Newark: University of Delaware, Disaster Resource Center.

Dynes, R. 2006. Social Capital: Dealing with Community Emergencies. *Homeland Security Affairs* 2 (2): 1–26.

Elliot, J., T. Haney, and P. Sams-Abiodun. 2010. Limits of Social Capital: Comparing Network Assistance in Two New Orleans Neighbors Devastated by Hurricane Katrina. *Sociological Quarterly* 51: 624–648.

Fabbe-Costes, N., M. Jahre, and C. Roussat. 2008. *The Contribution of Logistics Service Providers in Supply Chain Integration*. Avignon: RIRL.

Feigin, S., G. Owens, and F. Goodyear-Smith. 2014. Theories of Human Altruism: A Systematic Review. *Annals of Neuroscience and Psychology* 1 (1): 1–9.

FEMA (Federal Emergency Management Authority). 2000. *Hazards, Disasters and the US Emergency Management System: An Introduction, Section 6: Fundamentals of US Emergency Management*. Washington, DC: FEMA.

Fuchs, A., and K. Nils-Hendrik. 2012. Emergency Aid 2.0. https://papers.ssrn.com/sol3/papers.cfm?abstract_id=2519635. Last Accessed 4 July 2017.

GFDRR (Global Facility for Disaster Reduction and Recovery). 2013. *Post-disaster Needs Assessments, Vol. A, Guidelines*. New York: GFDRR.

GFDRR (Global Facility for Disaster Reduction and Recovery). n.d. https://www.gfdrr.org/post-disaster-needs-assessments. Last Accessed 15 Oct 2017.

Gillmann, N. 2010. *Interagency Coordination During Disaster: Strategic Choices for the UN, NGOs, and Other Humanitarian Actors in the Field*. Baden-Baden, Germany: Nomos Publishers.

GoB (Government of Bangladesh). 2008a. *Cyclone Sidr in Bangladesh: Damage, Loss and Assessment for Disaster Recovery and Reconstruction*. Dhaka: GoB.

GoB (Government of Bangladesh). 2008b. *Super Cyclone Sidr 2007: Impacts and Needs Strategies for Interventions*. Dhaka: Ministry of Food and Disaster Management.

GoN (Government of Nepal). 2015. *Nepal Earthquake 2015: Post Disaster Needs Assessment, Vol. A: Key Findings*. Kathmandu: National Planning Commission.

Good Humanitarian Donorship. 2003. Principles and Good Practice of Humanitarian Donorship. http://www.reliefweb.int/ghd/a%2023%20Principles%20EN-GHD19.10.04%20RED.doc. Last Accessed 15 July 2017.

Good Humanitarian Donorship Initiative. n.d. http://www.ennonline.net/fex/24/humanitarian. Last Accessed 15 July 2017.

Harmer, A., and E. Martin. 2010. *Diversity in Donorship: Field Lessons, Humanitarian Policy Group Research Report 30*. London, UK: Overseas Development Institute.

Harvey, P. 2007. *Cash-Based Responses in Emergencies*. London: Overseas Development Institute.

Hofmann, C.-A., and L. Hudson. 2009. Military Responses to Natural Disasters: Last Resort or Inevitable Trend? Humanitarian Practice Network. http://odihpn.org/magazine/military-responses-to-natural-disasters-last-resort-or-inevitable-trend/. Last Accessed 17 July 2017.

Holguin-Veras, J., and M. Jaller. 2012. Immediate Resource Requirements After Hurricane Katrina. *Natural Hazards Review* 15 (1): 1–12.

Holguin-Veras, J., M. Jaller, L.N. Van Wassenhove, N. Perez, and T. Wachtendorf. 2014. Material Convergence: Important and Understudied Disaster Phenomenon. *Natural Hazards Review* 15 (1): 1–12.

Humphries, V. 2013. Improving Humanitarian Coordination: Common Challenges and Lessons Learned from the Cluster Approach. *The Journal of Humanitarian Assistance* 30 April. http://sites.tufts.edu/jha/archives/1976. Last Accessed 15 July 2017.
IASC (Inter-Agency Standing Committee). 2006. Guidelines for Flash Appeal. https://www.humanitarianresponse.info/system/files/documents/files/7.5.1%20Flash%20Appeal.pdf. Last Accessed 12 July 2017.
IASC (Inter-Agency Standing Committee). 2015. *Nepal Earthquake 2015: Desk Review of Existing Information with Relevance to Mental Health & Psychosocial Support*. Kathmandu: IASC.
Islam, R., and G. Walkerden. 2014. How Bonding and Bridging Networks Contributes to Disaster Resilience and Recovery on the Bangladeshi Coast. *International Journal of Disaster Risk Reduction* 10: 281–291.
Ito, H., W. Wisetjindawat, and M. Yokomatsu. 2014. Improving the Operational Efficiency of Humanitarian Logistics in the Aftermath of a Large-Scale Disaster. *Journal of Integrated Disaster Risk Management* 4 (2): 142–155.
Johannessen, A., A. Rosemarin, A.G. Swartling, G. Han, G. Vulturius, and T.A. Stenström. 2013. *Linking Investment Decisions with Disaster Risk Reduction in Water Sanitation and Hygiene (WASH): The Role of the Public and Private Sectors, Potentials for Partnership and Social Learning*. Background Paper, Stockholm Environment Institute, Stockholm.
Kelman, I., and R. Dodds. 2009. Developing a Code of Ethics for Disaster Tourism. *International Journal of Mass Emergencies and Disasters* 27 (3): 414–431.
Kent, R.C. 1987. *The Anatomy of Disaster Relief: The International Network in Action*. London: Frances Printer.
Kragelund, P. 2010. The Potential Role of Non-Traditional Donors' Aid in Africa. ICTSD Issue Paper No. 11, International Centre for Trade and Sustainable Development, Geneva, Switzerland.
Lee, H.L. 2004. Creating Value Through Supply Chain Integration. *Supply Chain Management Review* 4 (4): 30–36.
Lee, H.L., V. Padmanabhan, and S. Whang. 1997. The Bullwhip Effect in Supply Chains. *Sloan Management Review* 42 (4): 54–62.
Lin, N. 2008. A Network Theory of Social Capital. In *The Handbook of Social Capital*, ed. D. Castiglione, J.W. Van Deth, and G. Wolleb, 50–69. New York: Oxford University Press.

Lindenberg, M., and C. Bryant. 2001. *Going Global: Transforming Relief and Development NGOs*. Bloomfield, CT: Kumarian Press.
Margesson, R., and M. Taft-Morales. 2010. *Haiti Earthquake: Crisis and Response*. Washington, DC: Congressional Research Service.
Middleton, N., and P. O'Keefe. 1998. *Disaster and Development: Politics of Humanitarian Aid*. London: Pluto Press.
Mileti, D.S. 1999. *Disasters by Design: A Reassessment of Natural Hazards in the United States*. Washington, DC: Joseph Henry Press.
Mitchell, J.K. 2006. The Primacy of Partnership: Scoping a New National Disaster Recovery Policy. *American Academy of Political and Social Science* 604: 228–255.
Montz, B.E., G.A. Tobib, and R.R. Hagelmann II. 2017. *Natural Hazards: Explanation and Integration*, 2nd ed. New York: The Guilford Press.
Moore, S., E. Eng, and M. Daniel. 2013. International NGOs and the Role of Network Centrality in Humanitarian Aid Operations: A Case Study of Coordination During the 2000 Mozambique Floods. *Disasters* 27 (4): 305–318.
Murphy, B.L. 2007. Locating Social Capital in Resilient Community-Level Emergency Management. *Natural Hazards* 4 (2): 297–315.
Nelan, M.M. 2016. The Social Construction of Donations: Agility, Adaptability, and Alignment as Success Determinants in Relief Supply Chains. A PhD dissertation, the University of Delaware, Newark, Delaware.
OCHA (Office for the Coordination of Humanitarian Affairs). 2007. *Oslo Guidelines: Guidelines on the Use of Foreign Military and Civil Defense Assets in Disaster Relief*. New York: UN.
OCHA (Office for the Coordination of Humanitarian Affairs). 2011. Financial Tracking System. http://fts.unocha.org/. Last Accessed 8 Aug 2017.
OCHA (Office for the Coordination of Humanitarian Affairs). 2012. OCHA on Message: Consolidated Appeal Process. http://reliefweb.int/sites/reliefweb.int/files/resources/ACF0BD6A42273A73852576AC0066571F-Full_Report.pdf. Last Accessed 12 July 2017.
OCHA (Office for the Coordination of Humanitarian Affairs). 2015a. Nepal Flash Appeal for the Response to the Nepal Earthquake, April–July 2015. http://reliefweb.int/sites/reliefweb.int/files/resources/nepal_flash_appeal.pdf. Last Accessed 14 July 2017.

OCHA (Office for the Coordination of Humanitarian Affairs). 2015b. Nepal Flash Appeal Revision: Nepal Earthquake, April–September 2015. https://reliefweb.int/sites/reliefweb.int/files/resources/nepal_earthquake_2015_revised_flash_appeal_june.pdf. Last Accessed 25 October 2017.

ODI (Overseas Development Institute). 2006. Good Humanitarian Donorship: June 2003 to March 2006. https://www.odi.org/projects/402-good-humanitarian-donorship. Last Accessed 15 July 2017.

Ozeguven, E.E., and K. Ozbay. 2013. A Secure and Efficient Inventory Management System for Disasters. *Transportation Research Part C: Emerging Technologies* 29: 171–196.

Patulny, R.V., and G.L.H. Svendsen. 2007. Exploring the Social Capital Grid: Bonding, Bridging, Qualitative, Quantitative. *International Journal of Sociology and Social policy* 27 (1/2): 32–51.

Paul, B.K. 2006. Disaster Relief Efforts: An Update. *Progress in Development Studies* 6 (3): 211–223.

Paul, B.K. 2011. *Environmental Hazards and Disasters: Contexts, Perspectives and Management*. Hoboken, NJ: Wiley-Blackwell.

Paul, B.K., B. Acharya, and K. Ghimire. 2017. Effectiveness of Earthquakes Relief Efforts in Nepal: Opinions of the Survivors. *Natural Hazard* 85 (2017): 1169–1188.

Powell, B. 2004. Where Should Your Money Go? *Time* 165 (2): 44–45.

ReliefWeb. 2008. *Glossary of Humanitarian Term*. New York.

Rey, F. 2001. The Complex Nature of Actors in Humanitarian Action and the Challenge of Coordination. In *Reflections on Humanitarian Action: Principles, Ethics and Contradictions*, ed. Humanitarian Studies Unit. London: TNI, Pluto Press.

Russell, T.E. 2005. *The Humanitarian Relief Supply Chain: Analysis of the 2004 South East Asia Earthquake and Tsunami*. Cambridge, MA: MIT.

Schaar, J. 2007. *Perspectives on Good Humanitarian Donorship,* 2 vols. Sheffield: HRI.

Schreurs, M.A. 2011. Improving Governance Structures for Natural Disasters Response: Lessons from the Indian Ocean Tsunami. In *The Indian Ocean Tsunami: The Global Response to a Natural Disasters*, ed. P.P. Karan and S.P. Subbiah, 262–280. Lexington: The University Press of Kentucky.

Simo, G., and A.L. Bies. 2007. The Role of Nonprofits in Disaster Response: An Expanded Model of Cross-Sector Collaboration. Special Issue, *Public Administration Review* 2007: 125–142.

Skjoett-Larsen, T. 1999. Third Party Logistics—From an Interorganizational Point of View. *International Journal of Physical Distribution and Logistics* 21: 112–127.

Small, D.A., and C. Cryder. 2016. Prosocial Consumer Behavior. *Current Opinion in Psychology* 10 (1): 107–111.

Smith, K., and R. Ward. 1998. *Floods: Physical Processes and Human Impacts*. New York: Wiley.

Stecklow, S. 2005. How Companies Help Charities and Cut Inventory. *Wall Street Journal*, Dec 10.

Stephenson, M. 2005. Making Humanitarian Relief Networks More Effective: Operational Coordination, Trust and Sense Making. *Disasters* 29 (4): 337–350.

Stephenson, M., and M.H. Schnitzer. 2006. Interorganizational Trust, Boundary Spanning, and Humanitarian Relief Coordination. *Nonprofit Management and Leadership* 17 (2): 211–233.

Street, A., and G. Parihar. 2007. The UN Cluster Approach in Pakistan Earthquake Response: An NGO Perspective. http://odihpn.org/magazine/the-un-cluster-approach-in-the-pakistan-earthquake-response-an-ngo-perspective/. Accessed 21 May 2017.

Szreter, S., and M. Woolcock. 2004. Health by Association? Social Capital, Social Theory, and the Political Economy of Public Health. *International Journal of Epidemiology* 3: 650–667.

Tamamoto, M. 2005. After the Tsunami, How Japan Can Lead. *The Far Eastern Economic Review* 168 (2): 10–18.

Taylor, A.J.W. 1989. *Disasters and Disaster Stress*. New York: AMS Press.

Taylor, G., A. Stoddard, A. Harmer, K. Haver, and P. Harvey. 2012. *The State of the Humanitarian System*. London: ALNAP.

Telford, J., and J. Cosgrave. 2007. The International Humanitarian System and the 2004 Indian Ocean Earthquake and Tsunamis. *Disasters* 31 (1): 1–28.

Thomas, A.S., and L.R. Kopczak. 2005. *From Logistics to Supply Chain Management: The Path Forward in the Humanitarian Sector*. Sebastopol, CA: Fritz Institute.

Tierney, K.L., M.K. Lindell, and R.W. Perry (eds.). 2001. *Facing the Unexpected: Disaster Preparedness and Response in the United States*. Washington, DC: Joseph Henry Press.

Toyasaka, F., and T. Wakolbinger. 2014. Impact of Earmarked Private Donations for Disaster Fundraising. *Annals of Operations Research* 221 (1): 427–447.

Troutman, E. 2015. What Happened to the Aid? Nepal Earthquake Response Echoes Haiti. *Aid Works*, June 19. http://aid.works/2015/06/nepal-haiti/. Last Accessed 16 Sept 2017.

Trunick, P.A. 2005. Special Report: Delivering Relief to Tsunami Victims. *Logistics Today*, 1–9.

UN (United Nations). 2010. Haiti Earthquake: 2010 Flash Appeal, Version 1.1. http://reliefweb.int/sites/reliefweb.int/files/resources/ACF0BD6A42273A73852576AC0066571F-Full_Report.pdf. Last Accessed 12 July 2017.

Van Wassenhove, L. 2006. Humanitarian and Logistics: Supply Chain Management in High Gear. *Journal of the Operational Research Society* 57: 475–489.

Versluis, A. 2014. Formal and Informal Material AID Following the 2010 Haiti Earthquake as Reported by Camp Dwellers. *Disasters* 39 (S1): S94–S109.

Wachtendorf, T., B. Brown, and J. Holguin-Veras. 2013. Catastrophe Characteristics and Their Impact on Critical Supply Chains: Problematizing Material Convergence and Management Following Hurricane Katrina. *Homeland Security and Emergency Management* 10 (2): 497–520.

Wakolbinger, T., and F. Toyasaki. 2011. The Impact of Funding Systems on Humanitarian Operations. In *Humanitarian Logistics: Meeting the Challengesof Preparing for and Responding to Disasters*, ed. M. Christopher and P. Tatham, 33–46. London: Kogan Page.

Wehrfritz, G., and J. Cochrane. 2005. Charity and Chaos. *Newsweek*, Jan 17, 30–32.

Whybark, D.C. 2007. Issues in Managing Disaster Relief Inventories. *International Journal of Production Economics* 108: 228–235.

Wiharta, S., H. Ahmad, J.-Y. Haine, J. Lofgren, and T. Randall. 2008. *The Effectiveness of Foreign Military Assets in Natural Disaster Response*. Solna, Sweden: Stockholm International Peace Institute.

Wishart, J.P. 2008. *Fostering Partnership in Humanitarian Aid and Disaster Relief*. Annapolis, MD: United States Naval Academy.

Woolcock, M. 2001. The Place of Social Capital in Understanding Social and Economic Outcomes. *Canadian Journal of Policy Research* 2 (1): 11–17.

World Bank. 2010. Damage, Loss, and Needs Assessment Guidance Notes, Volume 1, Design and Execution of a Damage, Loss, and Needs

Assessment. https://openknowledge.worldbank.org/handle/10986/19047. Last Accessed 28 May 2017.

Ysui, E. 2007. Community Vulnerability and Capacity in Post-disaster Recover: The Cases of Mano and Mikura Neighborhoods in the Wake of the 1995 Kobe Earthquake. Unpublished PhD dissertation, Vancouver, British Columbia, Canada.

3

Disaster Relief Provision

Among the crucial aspects of distributing disaster relief, inequitable or biased delivery, human rights violations, untimeliness, inappropriateness, and misuse and non-use of relief funds ultimately determine the success or failure of any single operation in response to an emergency; these are discussed in this chapter. Many hazard and disaster researchers (e.g., Hewitt 1983; Susman et al. 1983; Wisner et al. 2004) criticize emergency assistance to disaster victims because of these undesirable aspects of relief operations. Despite the criticism, certain aspects of disaster relief must be understood in association with several key paradigms and concepts of environmental hazards and disaster, particularly the political-ecological or political economy paradigm and the concepts of vulnerability and participatory exclusion. Despite several essential differences, most theories and approaches to responding to environmental hazards strongly argue that disaster relief distribution widely discriminates against marginal groups like the poor, elderly, children, illiterate, women, religious minorities, persons with disabilities, and immigrants/migrants.

Hazard research before the 1970s largely adopted the approach known as the "human ecological" tradition or behavioral perspective of

B. K. Paul, *Disaster Relief Aid*,
https://doi.org/10.1007/978-3-319-77282-0_3

hazards. Gilbert White developed this perspective and later expanded and popularized the perspective, along with his students, Richard Burton and Robert Kates (Emel and Peet 1989).[1] This tradition emphasizes natural hazards and disasters as a key component of the interaction between humans and environment, proposing the idea of a range of adjustments open to individuals facing the threat of hazards and emphasizing the importance of individual perception of, and response to, natural hazards (Burton et al. 1993; Wescoat 1987). Although White's influence spread internationally, in the 1970s, many hazard researchers later criticized this approach because they claim, its perspective is too narrow providing a singular focus on extreme natural events (Hewitt 1997; Wisner et al. 2004). It overexaggerates the role of the individual in hazardous and disaster situations, without paying enough attention to broader social and cultural factors (Mitchell and Cutter 1997). The shortcomings of the human ecological perspective have led to the development of the structural-political-economy approach to hazard research in the mid-1970s.

Researchers using the structural or radical approach claim that individual behavior during natural disasters and their aftermath cannot be understood without close analysis of the political, economic, and social structures on a broader scale. These structures not only affect human lives but also influence most other factors that determine individual vulnerability and behavior (e.g., Cutter and Finch 2008; Wisner et al. 2004). Followers of radical approach strongly believe that hazards arise from the societal injustices, deprivations, and inequalities (Blaikie et al. 1994) and thus have put much less emphasis on individual perception and decision making. They recognize the importance of societal and global-level forces in causing extreme natural events as well as biased distribution of disaster relief among victims of such events (Blaikie et al. 1994; Hewitt 1997). They posit that marginalized groups suffer more from natural disasters than non-marginalized groups because the former have less access to resources and hence differential susceptibility to natural hazards. Because of lack of resources and weak support networks, the marginalized and other vulnerable groups have always been excluded from environmental decision making (Pelling 1998, 1999; Jones and Murphy 2009).

Political ecologists are more concerned with issues like empowerment by gender, class, and caste; level of development; and international dependency in explaining the causal chain of vulnerability. In this sense, the political-ecological approach is similar to the concept of vulnerability, which has dominated the hazard literature for the last three decades or so. It also stresses that people of different backgrounds face similar risks from natural disasters, but their actual vulnerability lies in their socioeconomic conditions, civic and social empowerment, and access to resources and local support networks. Accordingly, members of marginalized and vulnerable groups are generally deprived of assistance in many societies during the response and recovery stages of disaster management cycle (e.g., Bolin and Stanford 1999; Cutter 2003; Wisner et al. 2004).

Of the different types of vulnerability, social vulnerability is closely associated with political-ecological approach to hazard and disaster research. Social vulnerability is influenced by income, educational levels, age, gender, caste, disability, and minority status. Social vulnerability suggests that both disasters and their effects are socially produced and is linked to larger-scale societal and global processes. It can also be linked to social-structural attributes of societies and communities. As with social vulnerability, the concept of participatory exclusion in the disaster context refers to systematic exclusion of significant sections of people in a society from ongoing disaster interventions, including relief operations (Agarwal 2001; Nadiruzzaman and Wrathall 2014, 2015). These people are aggregated as a marginalized group whose members live at the edge of a society, as insignificant and unimportant persons, and they are generally excluded from the disaster relief loop, systematically subjected to discrimination (Nadiruzzaman and Wrathall 2014). Geographically, they live in hazard-prone areas, which refer to their place vulnerability. Economically, they are poor and usually excluded from political processes (Robbins 2012).

In this chapter, we provide a backdrop for understanding the strengths and weaknesses of disaster aid disbursement as well as determinants of disaster relief at the country and individual/household levels. The information presented in this chapter should have immense value to emergency managers and planners, as well as many others who participate in emergency relief operations; the information should improve their capability and performance in the future disaster relief efforts as well as help them to implement better preparedness measures, responses, and recovery.

Biased Relief Distribution

The effectiveness of disaster aid to a large extent depends on how that aid is distributed, making sure that relief goods are delivered to those who need them the most. Uneven relief distribution can create conflicts and foster dissatisfaction within affected communities (Islam and Walkerden 2015). Several authors (e.g., Mallick et al. 2009; Aldrich 2010) maintain that uneven distribution weakens community resilience, leaves the poor more deprived, and delays post-disaster recovery. However, inequitable relief distribution is a perennial problem for almost all relief operations undertaken globally. Although this chronic problem is more prevalent in developing countries, developed countries are not completely immune. Inequitable relief distribution is frequently due to differences in religion, caste, socioeconomic status, local political involvement, gender, age, and region. For convenience in analysis, all types of biased distribution are broadly divided into four categories: socioeconomic and demographic, religious, political and personal gain, and region/area-specific biases.

Socioeconomic and Demographic Bias

This bias is consistent with the structural approach to environmental hazards and disasters as well as the concepts of vulnerability and participation exclusion. Unequal distribution of emergency aid has been widely reported for members of marginalized groups. As members of marginalized groups, the poor not only receive less amount of disaster relief but also receive low-quality and cheaper relief goods. Islam and Walkerden (2015) claim that, at the beginning of the relief phase (immediately after a disaster up to a week), relief agencies distribute lower quality items among the most vulnerable disaster survivors. In the next phase (from 1 week to a month), the agencies receive and distribute better quality of items among the moderately vulnerable people. Finally, (from a month to several months), the best quality items arrive in disaster-affected areas, and relief agencies distribute these goods among the less vulnerable people. They write, "Due to this relief distribution practice (and because of a household will not get the same relief items twice from the same organization), the best relief packages go to the least vulnerable people (comparatively better-off families) and the lowest-quality packages go to the most vulnerable people (the poorest)" (Islam and Walkerden 2015, 1717–1718).[2]

Male domination is a cultural or religious norm in many developing countries. Male members of disaster-affected households generally bring relief materials home from distribution points, which is a place where adult males often converge. Those who distribute relief goods are also frequently men. This environment at distribution points is undesirable for women, particularly for Muslim women who do not appear in front of unknown persons or talk to them. However, providing relief goods to women has many advantages. Sarah Bradshaw (2013, 96) explains it clearly: "… it is increasingly recognized that providing food to women is more likely to mean that the food stays with the family (rather than being sold for cash to spend on other goods) and to be distribute 'fairly' within the family." By fair distribution, Bradshaw means that women will bring relief food home, cook the food, distribute the cooked food among all family members, and they themselves will eat last. Once again, in many societies in developing countries, women eat last.

Bradshaw (2013) also pointed out problems associated with distribution of emergency aid to women. After the 2010 earthquake in Haiti, the World Food Program (WFP) initiated "women-only" food distribution in the capital of Port-au-Prince. Only women were authorized to enter the relief distribution points. Bradshaw (2013, 97) writes: "The environment surrounding distribution points entailed a real risk for many women, and there were reports of women being robbed of their food, and also of them being coerced to engage in sexual acts in order to receive extra food coupons." Delaney and Shrader (2000) noted another problem with distributing relief goods to women. They claim that after

Hurricane Mitch in 1998, women in Honduras and Nicaragua became busier and more involved than men in the daily work of processing relief goods. Note that not all women are equally discriminated in the disbursement of disaster relief. Women-headed households, widows, single women, and functionally illiterate women face more discrimination than other women.

Despite their special needs in non-emergency situations, in the response stage of the disaster management cycle, the elderly and physically disabled have many of the same needs as the general population. The stress of a disaster can cause significant challenges for anyone, but especially for seniors and people living with disabilities. After a major disaster, the elderly and disabled face serious hardship: They cannot sleep, confront considerable physical discomfort, and often stay hungry or miss familiar food. Unfortunately, they do not usually receive any special attention from relief agencies, particularly in developing countries. Relief workers provide no separate allocation of relief goods for them, and despite their inability to reach the relief distribution points, relief workers do not bring their emergency supplies to them.

After the 2004 Indian Ocean Tsunami (IOT), elderly and disabled people faced many problems in countries affected by the tsunami. In India, they became parents again for grandchildren who had lost their own patents to the tsunami. These deaths were also detrimental to the elderly themselves because their children were providing support to them before the tsunami. To support their grandchildren and themselves, they were forced to seek employment and confronted difficulties to adjust to the new situation (Paul 2007). Even when their own children survived, the elderly also faced other problems in some tsunami-affected countries. For example, some adult males living in camps in Sri Lanka consumed high level of alcohol, which had adversely affected their well-being. Additionally, elderly faced difficulty in obtaining medicine and resuming treatment for chronic illnesses in countries affected by the tsunami.[3] Children are also treated unequally in disaster relief distribution, particularly in developing countries. For example, in all countries affected by the 2004 IOT, relief organizations ignored the needs of children in distributing emergency aid. The items they distributed were all for adults people, a frustrating situation (Paul 2007). Children of poor families need supplementary feeding intervention because malnutrition increases among children of poor households during the post-disaster period (Hossain and Kolsteren 2003; Paul et al. 2012).

Religious Bias

Religion is a very sensitive issue for most of the people of the world. When distributors of relief goods are of different religions than the survivors, distributors are often perceived as biased by the survivors even if relief materials are distributed completely ignoring the religious affiliation of survivors. The reverse is also true. Often, a religious humanitarian mission offers disaster relief only to people belonging to that religion. For example, after the 1996 tornado in central Bangladesh, Ram Krishna (RK) Mission, a Hindu philanthropic organization, distributed relief only to Hindu survivors. Similarly, a Muslim charitable organization distributed relief goods only to Muslims (Paul 1998). After the 2015 earthquakes in Nepal, Paul et al. (2017) reported that some Buddhist priests in Sindhupalchok district provided relief goods only to Buddhists. Non-Christian survivors of the 2004 Indian Ocean Tsunami in Sri Lanka believed that several religious organizations were pursuing relief work with the sole objective of religious conversion (Nissanka 2006). In Sri Lanka, Christian relief organizations distributed relief goods to all religious groups. However, some Christians in Sri Lanka expected these organizations to provide emergency aid only to Christians.

Biased relief distribution is even seen among followers of same religion but belonging to different branches, sects, or castes. For example, during the devastating flooding in Pakistan in 2010, government authorities deliberately denied emergency aid to members of Ahmadiyya community, who are considered non-Muslims by most citizens of the country. Moreover, after their identity was revealed, they faced difficulty in remaining in relief camps set up for flood survivors (Paul 2011).

Similarly, after the 2015 earthquakes in Nepal, a report by Amnesty International claimed caste- and gender-based discrimination in providing relief for survivors (Specia 2015). The report further asserted that aid was not evenly distributed among women-headed households, the disabled, and the Dalit ("untouchable"—the lowest level of the Hindu caste system) population (Specia 2015). The Nepalese Dalit Civil Society (DNF 2015) confirmed this finding. The Dalits were also excluded from disaster relief in other parts of Nepal affected by earthquakes (SDI 2015). However, an empirical study conducted in two affected districts found no irregularities in distributing relief goods based on caste hierarchy (Paul et al. 2017). In reality, Maiti Nepal, a social NGO, distributed food items only to people belonging to the Dalit caste.

Human Rights Watch (2005) also observed caste-biased relief distribution among the 2004 IOT victims in affected areas of India. Although the caste system was abolished in India immediately after independence in 1947, it still prevails strongly in many parts of the country, particularly rural areas. However, the organization reported that public authorities as well as several aid agencies and local communities discriminated against Dalit communities in tsunami-affected areas. An on-site investigation by the National Campaign on Dalit Human Rights (NCDHR) found that some of the higher-caste fishing communities refused to share emergency shelter and rations with the Dalits. The NCDHR further accused that in Andhra Pradesh and Tamil Nadu states in India, coastal areas affected by the tsunami provided Dalits with less relief and financial assistance than non-Dalit survivors. Public authorities also waited to restore electricity and water supplies last in areas dominated by Dalit (Human Rights Watch 2005). A report by the International Dalit Solidarity Network (IDSN 2013) claimed that Dalits were systematically excluded from relief and recovery efforts in many other past disasters in South Asia, including the 2001 Gujarat Earthquake, 2008 flooding in Nepal and India, and in 2010 flood in Pakistan. This report stated that "Dalits face problems when trying to register in relief camps; they often receive poorer and fewer relief goods; they are exploited to remove corpses or debris; their losses are less visible as ownership is not formally recognised; and they are not consulted when needs are assessed" (Reliefweb 2013).

Political Bias and Personal Gain Bias

Disaster relief literature also provides ample evidence that emergency aid distribution is widely manipulated either for some political purpose and/or for personal gain. Supporters or activists of a political party in power usually receive more emergency aid than the supporters of opposition parties. For example, the 2004 IOT survivors in Sri Lanka complained that local relief officials provided emergency aid only to their supporters, some of whom were not even from tsunami-affected areas (Paul 2007). After the 2001 flood in Pakistan, Mustafa (2003) reported that personal connections with local political leaders and/or relief organizations meant those with connections were treated favorably in receiving additional relief materials. According to Pariyar (2015), and Sanderson and Ramalingam (2015), aid distribution was politically motivated in several parts in Nepal after the 2015 earthquakes. Observers in Nepal also accused several districts politicians of focusing on relief for their constituents rather than concentrating efforts on the worst affected victims and villages. Paul et al. (2017) claimed that not all NGOs provided disaster relief to all earthquake-affected areas in Nepal; one particular NGO's distribution was nearly all confined to only one or a few localities where they have been working on different projects for the past decade or more to improve the living conditions of the people.

In rural areas of developing countries, relief distribution is frequently influenced by local political dynamics and power relationships; both are linked to access to resources and livelihood opportunities. This led to the development of the idea of the "patron-client network" (see Bode 2002; Lewis and Hossain 2008), a mutually obligatory network between an individual (patron) and another person (client) who benefits from the patron's support or influence. Patrons have considerable influence in local communities because of their wealth and social status, while a client generally belongs to a marginalized group. In disaster relief operations, the client receives emergency items because of the authority of the patron in the community or society. Traditionally, local power elitism is inherited and historically confined to a few families. However, in recent decades, traditional local leadership has been challenged by newly emerging power networks, "which evolve through their affiliation with the wider institutional networks of partisan politics, NGOs, businesses and so on" (Nadiruzzaman and Wrathall 2014, 5).

Based on field observations and informal discussions with survivors of Cyclone Sidr, which hit the Bangladesh coast on November 15, 2007, Nadiruzzaman and Wrathall (2014) concluded that disaster relief was not allocated according to need, but through kinships and affiliations within power networks. While collecting data on the 2007 Cyclone Sidr disaster relief from two coastal villages in Bangladesh, Islam and Walkerden (2015) reported that NGOs generally favored their own borrowers as they provided relief and livelihood support (e.g., distribution of livestock, poultry, nets, and boats) among Cyclone Sidr victims (also see Paul and Hossain 2013). They further claimed that some NGO officials took bribes when they distributed livelihood support among Sidr survivors: "… some people received cows/calves by paying Tk. 5000 in bribes, and some people's names were cut from the distribution list because of their inability to pay the bribes" (Islam and Walkerden 2015, 1717).[4] In addition, some NGOs distributed boats for Tk. 200,000 ($2597), of which 60% was pocketed by NGO staff members. One international NGO cancelled a partnership with a local NGO because of its corrupt distribution of relief. Additionally, Transparency International Bangladesh reported that 7.2% of households paid bribes to NGO officials, on average Tk. 549 ($8) for services received from June 2009 to May 2010 (TIB 2011).

Paul and Rahman (2006) reported that most households in cyclone-affected islands in Bangladesh received no emergency relief because of the extensive corruption among NGO workers and officials (also see Mahmud and Prowse 2012). It seems that bribery and corruption was relatively widespread in the distribution of Cyclone Sidr emergency relief. According to a local key informant, one NGO worker lost his job after Sidr because of taking bribes (Islam and Walkerden 2015).

Region/Area-Specific Bias

Providing emergency relief only to specific regions or areas has also been reported in many disaster relief operations. It occurred in Pakistan after the country hit by an earthquake in 2005. Even Hurricane Katrina survivors of New Orleans and Louisiana perceived that they received less emergency aid than survivors living in Alabama and Mississippi (Paul 2011). The 2004 India Ocean Tsunami survivors of Tamil Nadu, India, reported that the government and other relief agencies exaggerated the amount of damage in some areas and underestimated the damage in other areas. As a result, survivors of many severely affected areas did not receive adequate relief, while others in far less dire straits received plenty of emergency aid (P'Rayan 2005). Discrimination in distributing disaster relief in Tamil Nadu was also reported by occupational groups. In this state, farmers received less in-kind relief goods and cash than fishermen. Sarah Bradshaw (2013) echoed the unfair relief distribution. She writes: "Each new 'disaster' brings new reports of some villages receiving funding from three or four organisations while others receiving nothing, or the same goods being provided by a number of different organisations, such as housing and latrines, while potable water is supplied by no one" (Bradshaw 2013, 98). In Haiti, researchers (e.g., Kirsch et al. 2012; Versluis 2014) reported that 2010 earthquake survivors who were living in camps received more emergency aid than survivors who did not.

Timeliness of Relief Distribution

The timely delivery of emergency goods to the disaster-affected population is crucial to the disaster relief supply chain because it decreases fatalities and reduces survivors suffering and distress. The overarching objective of disaster relief is always linked to how quickly and conveniently the emergency aid reaches affected people. This led to the development of the term "last mile relief distribution" in supply chain literature. It represents the final stage of a humanitarian relief chain and refers to moving relief items from local distribution centers (LDCs) to the people affected by the disaster (Balcik et al. 2008). This implies that necessary emergency supplies must be stored at LDCs before disaster strikes which will facilitate quick distribution of basic needs to disaster survivors (also see Chapter 2).

However, timely distribution is also needed to prevent chaos in the immediate aftermath of an extreme event (Nagurney et al. 2015; Nagurney and Nagurney 2016). Other than medical supplies, if disaster survivor emergency needs are not met within 72 hours of the event, they become desperate, seeking any possible means to acquire food, drinking water, and other necessary emergency materials (see Walton et al. 2011; Fugate 2012). Slow or late delivery of disaster relief therefore often leads to distressed survivors taking to the streets in protest or attacking relief trucks. Such attacks were reported for recent disasters like the 2010 Haiti Earthquake, 2013 Typhoon Haiyan in the Philippines, 2015 Nepal Earthquakes, and 2017 Hurricane Maria in Puerto Rico (Hebblethwaite 2011; Nagurncy et al. 2015; Paul et al. 2017).

Lack of timely relief distribution may be due to lack of preparation or coordination among the appropriate authorities or among private and public agencies involved in disaster relief efforts. This also implies that the government of a country struck by disaster is not efficient enough to conduct relief operation quickly because of rigid bureaucracy, lack of qualified persons with past experience of disaster relief, absence of trained personnel, political instability or dysfunctional government, government insensitivity, and civil unrest. Remoteness, inaccessibility, and terrain may also cause delays in relief operations as was the case for Puerto Rico after it was hit by Category 4 Hurricane Maria on September 20, 2017, as well as for Nepal after two major earthquakes in 2015.

The federal response to Hurricane Maria's devastating toll on Puerto Rico was delayed by at least a week. Although the blame for this slow response was attributed to destruction of infrastructure, physical inaccessibility of remote affected areas, and shortage of fuel and drivers, survivors of the hurricane desperately needed food, water, medicine, and electricity. Additionally, the hurricane knocked out power and communications across the island and damaged air-traffic control towers. In contrast, the US response to the 2010 Haiti Earthquake was quick and remarkable. The US military was mobilized and sent to Haiti within 24 hours of the earthquake, but it took more than one week for them to arrive in Puerto Rico after Hurricane Maria. On the Haiti response, Davis et al. (2017) write: "Within two days, the Pentagon had 8,000 American troops en route. Within two weeks, 33 US military ships and 22,000 troops had arrived. More than 300 military helicopters buzzed overhead, delivering millions of pounds of food and water." Strikingly, eight days after Hurricane Maria ripped across Puerto Rico, just 4400 service members as well as about 1000 Coast Guard members were present, participating in federal relief operations to assist Hurricane Maria survivors. About 40 US military helicopters were used to deliver food and water to the 3.4 million residents of Puerto Rico, along with 10 Coast Guard helicopters (Davis et al. 2017).

In 2015, two major earthquakes struck Nepal's northern mountainous and central hilly regions, which then suffered delays in receiving relief. The earthquakes severely damaged roads and infrastructure, making it nearly impossible to transport relief goods to the earthquake-affected sites without first clearing and repairing major roads. Other serious barriers to quickly clearing roads were severe shortages of earthmovers, cranes, supply trucks, fuel, and drivers (Paul et al. 2017). Shortages of trucks and drivers also delayed by over a fortnight transport of essential relief goods after the 2004 flood in Kashmir (Ali and Ince 2017). As indicated, shortages of truck drivers and fuel also caused delays of more than a week in transporting relief goods from San Juan, Puerto Rico, to remote locations severely affected by Hurricane Maria in 2017. However, in Nepal, a series of high-magnitude aftershocks added to the problems with the road conditions. These aftershocks caused landslides, which blocked many roads and further delayed assistance to the farthest reaches of the country (Paul et al. 2017). Nepal also did not have adequate numbers of helicopters to airdrop relief supplies in remote areas affected by the earthquakes. Initially, the government used only the three helicopters owned by the United Nations. These three helicopters were far too few for the vast area affected by the earthquakes. Helicopters were also vulnerable to the mountain and hilly terrain of the country (Troutman 2015).[5] Extreme fuel shortages also left relief organizations unable to reach remote communities.

However, instead of blaming geographical conditions and severely damaged infrastructure for delays in providing relief, some critics accused the government and its bureaucracy of causing the delay (Weesjes 2015). The government imposed a number of rules and regulations for relief goods entering the country from foreign sources. All goods went through customs inspections at the Tribhuvan International Airport. Beginning on 26 May, the Nepalese government started to charge import taxes on almost all goods coming from outside the country other than medicine (Troutman 2015; Weesjes 2015). This caused foreign relief supplies to pile up at the airport, which, in turn, delayed the distribution of emergency aid (Paul et al. 2017). In this case, however, the people of Nepal seemed to whole heartedly support the government's position on imposing import taxes and reviewing emergency items coming from abroad (Arora 2015). They expressed their unwillingness "to accept emergency aid at the cost of the nation's sense of sovereignty and self-respect" (Paul et al. 2017, 1184).

Customs inspection also delayed delivery of disaster relief in Sri Lanka after its coastal region was hit by the tsunami in 2004. At that time, the country was engaged in a full-scale civil war with the Tamil of the northeastern part of the country demanding a separate Tamil state. Militant fighters of Tamil Liberation Army (TLA) were receiving arms from Tamil-speaking people living abroad. As a result, Sri Lankan customs officers were checking all incoming boxes containing relief items from foreign countries. This, coupled with clashes among political parties, caused considerable delay in getting relief items to survivors of the tsunami, particularly in Tamil-held territory. A similar situation was also reported in the Aceh province in Indonesia where the military wing of the Free Aceh Movement (*Gerakan Aceh Merdeka* [GAM]) was fighting government forces (Paul 2007, 2011).

Several international aid agencies, as they tried to provide emergency supplies to survivors of the 2004 IOT in Indonesia, were frustrated by the inefficiencies of government officials. For example, the Australian Red Cross sent a huge quantity of bottled water by ship for the tsunami survivors in the Aceh province of Indonesia. Almost five months later, not a single bottle of water had reached survivors. Instead, these water bottles remained undistributed at the docks in the port of Medan in

Sumatra. Indonesian customs officials claimed that a shortage of trucks and slow accounting processes were responsible for this long delay in distributing not just bottled water, but more than 4100 cargo containers of medical supplies, hundreds of tons of food, tents, clothing, and food (Donnan 2005). In fact, one main problem in Indonesia was that relief supplies stacked up of at the airports at Banda Aceh and Medan. Lack of unloading equipment caused considerable delay in transporting relief goods from these airports to distribution points. Additionally, both the central and the Aceh provincial governments had little infrastructure to help in quickly dispersing relief materials (Margesson 2005).

Developing countries are not alone in experiencing delays in providing relief items to areas affected by disaster impact; developed countries often see similar delays. The survivors of 2004 Hurricane Katrina did not receive emergency supplies promptly. The local and state governments blamed laid this delay squarely on the federal government. In fact, however, when natural disasters strike in the USA, state and local governments have the primary responsibility for disaster relief. In the USA, if a state suffers a natural disaster, the state government must request the federal government to declare the affected area a disaster area. Until that declaration, the federal government is not responsible for emergency relief. Louisiana was one of the several states severely affected by Katrina. Its governor, Kathleen Blanco, failed to request federal assistance in a timely manner (see Christophe and Adams 2016).

Misuse/Non-Use, Diversion, or Double Counting of Disaster Relief

Selling of in-kind emergency aid on the open to market has been reported during post-disaster periods in many countries. This type of selling was probably relatively widespread during the 2004 IOT relief efforts. Reliefweb (2005) reported tsunami victims in Aceh province, Indonesia, selling canned sardines and in some tsunami-affected coastal areas of Sri Lanka, many male heads of households diverted cash intended to buy emergency items like food and instead bought alcohol and drugs. In some areas, men forced women to give up excess food received as relief goods and used the food to buy alcohol. This led to increase of domestic violence in tsunami-affected areas (Fisher 2005). In another example, the Bangladesh government and other international agencies provided grants to build houses for the survivors of Cyclone Sidr, which hit the country's coastal area in 2007, but many recipients spent their grant money to food or winter clothing for children (Paul and Rahman 2013).

Part of the problem is the unfamiliarity of some relief items. Disaster survivors often do not use items received from relief agencies because the items are unknown to them or they do not know how to use the items. For example, after the 2005 earthquake in Pakistan, foreign relief organizations distributed toothpaste in several communities in Kohistan. Many of the people of these remote communities had never seen toothpaste in their lives, so they ate it, thinking it was a foreign sweet (IFRC 2006). In other communities, women were provided sanitary pads when they had never used pads before. In yet in other areas, foreign relief organizations distributed kerosene heaters among the earthquake survivors when kerosene oil was unavailable or too expensive to buy; agencies discontinued distribution of these heaters (IFRC 2006). Organizations must provide survivors relief items, but they also need to provide information about the use of those items. This issue is discussed in greater detail in Chapter 5.

Disaster survivors are not the only people to misuse, mishandle, or fail to use disaster aid. Some humanitarian agencies often misuse the money they collect for a particular disaster, diverting the funds for other purposes. Millions of dollars were raised by several organizations, including the American Red Cross, for rebuilding Haiti after the 2010 earthquake, but only about 9000 houses were ever built. According to Troutman (2015), "Most of the money never got to the people in need" (Troutman 2015). Generally, not all donated cash benefits disaster victims. A portion of this cash, usually 10%, covers expenses associated with fund-raising and/or as administrative costs.[6] After the Nepal Earthquakes in 2015, the Disaster Accountability Project, a non-profit organization that promotes transparency and oversight in disaster management systems, conducted a survey of organizations soliciting donations for Nepal. The findings revealed that several agencies collecting money for relief in Nepal were diverting the funds to other projects, whose details were not specified (Zakaria 2015). For example, ShelterBox, a charity based in Cornwall, England, collected donations for Nepal Earthquake victims, but Troutman (2015) claims that not all donations ended up in Nepal. In its print donation page, the Shelterbox states that all donations will go to the organization's general fund and will be spent as necessity dictates, a strategy used by many humanitarian organizations.

When several organizations collaborate in providing disaster aid, moreover, they often take credit for the same activity, resulting in double or triple counts of international responses. Troutman (2015) provided several examples of such counting during the relief operations after the 2015 Nepal Earthquakes. She reported that two relief organizations, Medair and Mission East, collaborated to deliver emergency assistance to earthquake survivors in Nepal. Medair received funding for shelter, and Mission East for water and sanitation. Troutman (2015) claims that both organizations took credit for both activities, resulting in a double count of the same aid.

Sometimes there is even triple or quadruple claims for the same relief efforts. According to the United Nations Office of the Coordination of Humanitarian Affairs and a report from humanitarian shelter officials, as of mid-June, 2015, some 762,000 earthquake survivors of Nepal were provided with either a tent or a tarp. However, the international aid community reported distribution of those two items to 3 million people (Troutman 2015). Emily Troutman, a journalist who investigated this discrepancy, concluded that several international aid agencies took credit for the same aid. She examined the Web sites of 45 aid organizations operating in Nepal since the earthquakes and reached to the above conclusion. She further claims that the US Agency for International Development (USAID) provided tarps to the International Organization for Migration (IOM). The latter organization then gave the tarps to a French relief organization, the Agency for Technical Cooperation and Development (ACTED). With the help of local workers, ACTED distributed tarps to the earthquake survivors. However, all three organizations took credit (Troutman 2015).

The misuse of relief is not confined to the recipients themselves or relief organizations. All entities (e.g., governments of affected countries, U.N. agencies, and NGOs) involved in 2004 IOT relief efforts established a range of mechanisms to ensure transparency and accountability to donors, the general public, and the affected populations. The governments of these countries were supposed to set up special audit mechanisms and anti-corruption conventions (U.N. n.d.). Despite this, available literature suggests states and national governments in Indonesia, India, and Sri Lanka diverted or misused relief funds intended to help survivors of the 2004 IOT. In Indonesia, some of the tsunami relief funds may have fallen into the hands of fundamentalist Islamic groups like Mujahideen Kompak and Majelis Mujahideen Indonesia or MMI, which have been involved in terrorist activities within the country and abroad (Paul 2007). A government audit in Indonesia found $5 million in relief funds lost to corruption in the first few weeks after the 2004 IOT (*Washington Post* 2006).

One government report in India claims that $10 million earmarked for tsunami relief in India was siphoned off to cover existing liabilities and administrative overhead. In fact, tsunami survivors in India had been "adversely affected" because $4 million in relief funds was never used. This unused money was simply returned to the central government (Kumar and Kala 2006).

A report from the National Audit Office (NAO) in Britain noted that as late as March 2006, some tsunami aid fund raised in Britain might still have been unused. The NAO further reported that "it is unclear how much of the original $600 million has been spent, and how much is in the bank accounts of various entities" (Paul 2007, 475). The report praised the Department for International Development's initial response to the disaster but expressed concern about the $100 million it allocated to U.N. agencies, NGOs, and charities (*BBC News* 2005).

The situation was no different after the 2010 Haiti Earthquake. After this devastating event, the American Red Cross raised almost half a billion dollars. A review of hundreds of pages of Red Cross internal documents and emails, as well as information from a dozen current and former officials, NPR, and ProPublica, found a string of poorly managed projects, questionable spending, and dubious claims of success.

The Red Cross claims it provided homes to more than 130,000 people, but the organization built only six permanent homes (Elliot and Sullivan 2015). In another case, because of alleged fraud in Aceh, Indonesia, Oxfam Great Britain took disciplinary action against 22 of its local staff members. This clearly implies that Oxfam took the issue seriously and will not tolerate such fraud in the future.

The U.N. also alleged misuse of the fund established to assist in relief after the 2004 IOT. This organization not only launched a Flash Appeal for immediate assistance but also took the responsibility to coordinate all tsunami relief efforts in the early post-disaster period, creating an online, global Financial Tracking System for the response phase (U.N. n.d.). However, some critiques (e.g., Donnan 2005) have questioned the U.N. use of this relief fund. A two-month-long investigation conducted by the London-based *Financial Times* found serious problems in the use of tsunami emergency aid. According to the *Financial Times* inquiry, the U.N. used relatively large amounts of available relief funds to organize international conferences and paid as much as $10,000 a month to international aid consultants. The U.N. spent 33% of the tsunami aid money on administrative costs; Oxfam spent just 10% on such costs (Donnan 2005). The U.N. often spent three times more on travel for its staff members, accommodations in hotels, and attendance in tsunami-related conferences than did private charities (Donnan 2005). Hazard researchers suggest that the U.N. organized too many conferences on the 2004 tsunami. The U.N. even paid very high salaries for its international consultants (Paul 2007).

Other Issues

As if the previous irregularities were not enough for disaster survivors to endure, they face other problems as well. These problems can be subdivided into two interrelated groups: human rights violations and health/sanitary issues. These two issues are interrelated because problems with latrines or temporary shelters immediately after a disaster can be a factor in violence against women. This section explores the interrelation between health/sanitary issues and human rights violations. Note that gender discrimination, which was discussed earlier in this chapter, appropriately falls under human rights violations.

During the response phase of disaster management cycle, human rights violations are very common, particularly in developing countries. As noted, these violations generally occur in the form of gender-based discrimination in disaster relief distribution as well as sexual violence against women and girls. The first component of human violations was discussed in an earlier section of this chapter. This subsection will be devoted to the second component. The difficulties disaster survivors face early in the post-disaster period not only differ by gender, but also by type of disaster. Women and girls are the victims of sexual abuse at relief distribution sites as well as in temporary shelters or camps built for disaster survivors. In many developing countries, rural areas suffer a serious lack of sanitary facilities. Most males and many females in these areas urinate and defecate in open spaces, a constant challenge for women, a challenge that becomes worse during or after disasters like river flooding and flooding associated with storm surges.

Storm surges in coastal areas contaminate rural drinking water sources (e.g., mud wells, tube wells, and ponds) with saline water and debris. In coastal urban centers, power outages affect water supplies in areas with piped water. This lack of pure drinking water helps spread waterborne and infectious diseases among the population (especially children and pregnant women). Moreover, women and adolescent girls in rural coastal areas must walk great distances to fetch drinking water. After waiting in long lines for water (sometimes little more than a pitcher), they are at risk of assault and rape as they travel home.

Contaminated drinking water sources and damage to household latrines, mostly uncovered pit latrines, were reported after the two most recent tropical cyclones to strike coastal Bangladesh: Cyclone Sidr in 2007 and Cyclone Aila in 2009 (Table 3.1).[7] Few latrines have septic tanks. Both cyclones caused considerable damage to sanitation facilities, particularly in the worst affected areas. Most of the damage was concentrated in rural areas, and in the worst affected areas, 70% of the slab latrines were completely destroyed (GoB 2008). Aila made landfall at high tide, which resulted in deadly storm surges up to 22 ft. (6.7 m) high. The force of the surge was enough to breach and destroy 5000

Table 3.1 Water supply and sanitation damage caused by Cyclones Sidr and Aila, Bangladesh

Facility (number)	Cyclone Sidr[a]	Cyclone Aila[b]
Tube wells	11,612	594
Ponds	7155	3195
Latrines	55,000	59,400

[a]Most of the damage occurred in four severely affected coastal districts (second largest administrative unit in Bangladesh; each district contains about 2.5 million people)
[b]Most of the damage occurred in two worst affected districts
Sources Prepared by the author based on information provided by GoB (2008, p. 27) and International Agencies (2009, p. ii)

miles (7500 km) of coastal embankment networks built in the 1960s to protect coastal residents from cyclones and tidal surges. The surge also dropped a thick layer of mud sludge in southwestern and central coasts of Bangladesh and caused an inundation of vast areas of land with saline water. Some areas remained under water for months (Walton-Ellery 2009).

Flooding caused by storm surges associated with Cyclone Aila forced many people, both male and female, in affected areas to resort to open defection. At that time, coastal residents were financially unable to install latrines. This caused serious problems for women and adolescent girls. They were forced to either share latrines with other people or travel to find remote places for open defection. Sharing a latrine among 10–15 households created other problems, particularly odors and difficulty in siting on the latrine slab without falling.

At temporary shelters, women and adolescent girls lived in constant fear, threatened by overcrowding and a lack of privacy. These shelters lack adequate lighting, toilets, and running water. Most temporary shelters had a single room without the space or privacy to change their clothes. The materials used to make partitions between two adjacent rooms further encroached on women's privacy.

Even when they lived in their home compound in tents, women faced similar challenges. For example, after the 2015 Nepal Earthquakes, survivors received tents as donations from foreign countries. However, the tents were inadequate for large households. These tents protected survivors from rain and storms reasonably well during the monsoon season but failed to protect them from winter cold. Moreover, dew from the tent roofs, caused by very low roofs, was a problem, and many elderly, women, and children died from hypothermia (Paul et al. 2017). Again, the tents were small, lacking a separate room for women and girls, and they felt uneasy changing their clothing in such close proximity to male family members (Daniel and Mahr 2015).

However, available information suggests that women confronted many challenges in the temporary shelters constructed for disaster victims in several countries affected by the 2004 Indian Ocean Tsunami-affected countries. One such challenge was associated with latrines. Many camps were without adequate latrines, and most were located some distance away from the camps. This presented a challenge for the women, particularly at night because lighting was limited. Additionally, many latrines were hastily and/or poorly constructed, leaving gaps in the door, so women preferred to use them at night. Such latrines were also not accessible to elderly, especially at night, because of their location and the lack of adequate lighting (U.N. 2006). In India and Sri Lanka, women complained that latrines built close to the temporary shelters lacked water and lights during the dark hours, and that they had to walk long distances to use these facilities (Paul 2007). These inadequacies led to a large number of sexual harassment and rape cases reported in these camps, particularly in Aceh, Indonesia. The situation was no different in the two large temporary shelters (the Superdome and the New Orleans Convention Center) created for survivors of Hurricane Katrina in New Orleans, Louisiana. Rapes and sexual abuse were reported in both shelters, which are discussed in the next chapter.

Additionally, in Tamil Nadu, India, where temporary structures were built with tar sheeting after the tsunami, women were left with the difficult choice of sleeping outside with other members of the community, which put them at risk of abuse, or risking heat-related health problems by sleeping inside these suffocating shelters (Paul 2007).

Fortunately, water, sanitation, and hygiene (WaSH or WASH) programs have been incorporated as one element of a cluster introduced by OCAA in 2005. This program has also identified several international development agencies like USAID, UNICEF, and WHO with the capacity to improve health, gender equality, and other important issues of international development. A unique feature of WaSH is to provide males, females, and adolescents with hygiene education to help promote better health. In Bangladesh after Cyclone Aila, this program ensured separate sanitation facilities for girls and female teachers at secondary schools. As a part of a strong gender orientation, this program emphasizes the participation of women in all decisions related to implementing WaSH program. To supplement WaSH programs, the Government of Bangladesh and many NGOs have undertaken a number of projects to improve health: installing tubewells with raised platforms, building low-cost water treatment plants and low-cost rainwater harvesting plants, and installing low-cost latrines in coastal Bangladesh after Cyclone Aila in 2009 (MoFDM 2009).

Effectiveness of Disaster Relief Distribution

Because all these problems associated with disaster relief distribution, distribution performance must be critically evaluated after each major disaster. As noted, many public agencies, humanitarian organizations, and others are involved in delivering emergency aid to disaster victims. For various social, political, geographical, economic, and cultural factors, each relief operation confronts unique challenges. This means, every disaster differs, and every disaster requires different responses in relief distribution. A careful evaluation of relief operation can provide valuable insights and lessons for those who disburse emergency assistance. These insights and lessons can assist both public and other relevant organizations in developing efficient and orderly relief operation plans for survivors of future natural disasters.

All distributors together can be evaluated at once for their overall effectiveness in distributing emergency aid. Alternatively, a comparative approach, such as nongovernmental organizations (NGOs) Versus government organizations (GOs), can also examine the effectiveness of disaster aid distribution. Such a comparison might be useful if the official response is widely criticized for inefficiency that fails to deliver timely relief at the disaster site (Holguin-Veras et al. 2007; Holguin-Veras and Jaller 2012). Disaster literature suggests that, until recently, NGO responses were more efficient, effective, and impartial in disbursing critical emergency assistance to disaster survivors than most governmental agencies, which are controlled by elite groups and hard-core bureaucrats, also known as command-and-control disaster relief operations (e.g., Paul 2003, 2006, 2007). This implies that GOs systematically discriminate against marginal households and socially vulnerable populations in providing disaster aid. A comparative approach among linking networks, bonding social networks, and bridging networks can also be pursued to examine the effectiveness of disaster relief distribution.

Ideally, evaluating the effectiveness of emergency aid should use the opinions of disaster survivors or others not directly involved in delivering aid but with clear knowledge about a specific relief operation. Both qualitative and quantitative approaches can be used to examine emergency aid effectiveness. In either case, several salient features of emergency relief disbursement should be considered in developing a composite index of effectiveness. While little effort has gone into developing such an index, Paul et al. (2017) recently developed a spatial level additive composite index of effectiveness of disaster aid distribution in Nepal after the two major earthquakes in 2015. This index was based on respondent opinions of the appropriateness, adequacy, timeliness, and fairness of aid distribution. According Paul et al. (2017), these four components/dimensions captured all salient features of effective emergency aid. A 1–5 Likert scale, anchored at 1 signifying the lowest opinion and 5 signifying the best, was used to record how respondents rated each of the four components measuring emergency relief efforts for each selected study unit and the entire study area.[8]

Using the answers from 272 respondents selected from two study districts (Sindhupalchok and Lalitpur), Paul et al. (2017) calculated the overall effectiveness of relief efforts first by adding the scores of all four components and then by taking the mean of the scores.[9] Because the two study districts differed in the extent of damage, distance from the capital of Nepal, and respondent characteristics, a comparative approach was used to examine how effectively aid was distributed to respondents in both Sindhupalchok and Lalitpur districts. Table 3.2 presents scores for the four selected components of emergency relief

Table 3.2 Mean score for four selected components of emergency relief efforts by selected districts in Nepal after the 2015 earthquakes[a]

Component	Study district		
	Sindhupalchok	Lalitpur	Total
Appropriateness	2.94	2.51	2.75
Adequacy	2.34	2.54	2.43
Timeliness	2.03	2.46	2.22
Fairness	2.75	2.67	2.71
Aggregate	2.52	2.55	2.54

[a]Mean scores range from 1 to 4
Source Prepared by the author based on Paul et al. (2017, p. 1183)

efforts. For the first component (i.e., appropriateness), 1 is allocated for "not needed at all," 2 for "not needed much," 3 for "more or less needed," and 4 for "urgently needed." For this and the remaining three components, no one marked the mid-value (3 which represented no opinion) of the Likert scale, so this value was dropped. The average score is 2.75 out of a total of four, showing that respondents were more or less satisfied with the materials they received as relief goods (Table 3.2).

For the second component, respondents were asked whether the relief materials they received were adequate for their need. About 54% of the 272 respondents reported the amount of aid they received was inadequate. The remaining 46% responded the aid was either somewhat adequate or inadequate (Paul et al. 2017). The overall adequacy score is 2.43, lower than the appropriateness score (Table 3.2). For timeliness of disbursal of emergency assistance, nearly two-thirds of the respondents stated that distribution was delayed; only 4% respondents reported that distribution was timely, and another 31% considered it was somewhat timely (Paul et al. 2017). On fairness of distribution of relief goods, 67% of all respondents considered distribution was mostly and/or completely fair. Note that apart from individuals, fairness in relief distribution also ensures equal or equitable distribution among communities and locations (Bradshaw 2013). Here, the term refers to differences between individuals and between the two study districts.

Table 3.2 shows the average scores for all four components measuring effectiveness of disaster relief aid in Nepal were higher than two for both districts as well as for the study area as a whole. The same is true for the overall effectiveness scores. This indicates that the respondents in both districts were satisfied with the distribution of earthquake relief, which, in turn, suggests that the disaster relief distribution was more or less unbiased in Nepal after the two major earthquakes in 2015.

As noted in Chapter 1, emergency assistance does not compensate for all losses and damage incurred during a disaster. However, from the financial point of view, disaster survivors often receive more than what they lost. Paul and Leven (2002) reported that several households in Hoisington, Kansas, USA, received more assistance than what they lost in 2001 in a tornado that hit this small city of about 1500 people. While conducting empirical research in Nepal in 2015 and 2016, Paul suspected that some households also received or will receive more than what they lost due in the 2015 earthquakes. Although nothing can compensate for loss of life, the Nepal government distributed 30,000 Nepalese rupees (NPR), or the equivalent of $300.00 for funeral costs for those households who lost a family member during the 2015 earthquakes (Asia Foundation 2016). The earthquake survivors also received a cash grant determined on the basis of the extent of damage incurred during the event. Families whose houses were destroyed received NPR 15,000 ($150.00) to build temporary shelter; those whose houses were partly damaged received NPR 3000 ($30.00). The Asia Foundation (2016) reported households damaged in medium and higher impact districts received NPR 15,000 ($150.00) as early cash assistance, while those in less affected districts received less.

Other than this early cash grant, earthquake survivors also received winter relief grants of NPR 10,000 ($100.00) to buy warm clothing (Paul et al. 2017). The survivors complained that the winter grant was inadequate and arrived too late to be helpful. Under the Nepal Rural Housing Reconstruction Program (PHRP), the Nepalese government originally planned to provide NPR 200,000 ($2000) to rebuild destroyed houses in rural areas. This amount was scheduled to be distributed in three installments of NPR 50,000 ($500.00), NPR 80,000 ($800.00), and NPR 70,000 ($700.00). Later, the government increased the grant to NPR 300,000 ($3000) (Asia Foundation 2016). In addition to government grants, earthquake survivors received cash from some NGOs along with in-kind donations. If we add the total monetary value of cash and in-kind donations, a considerable number of survivors received more than they lost. Unpublished field data collected by a research team organized by the NSF earthquake project indicated that on average damage in monetary value was NPR 856,764 ($857.00) (Paul et al. 2017).

As noted, various sources provide emergency aid to disaster victims. In monetary terms, this aid differs substantially from one source to another. Three decades of field work in the USA, Bangladesh, and Nepal suggest that the public sources provide the most aid, but disaster victims have always been dissatisfied with the amount of disaster assistance provided by these sources. They seem to be very happy with private sources who often contribute much less than the public sources. For example, Paul and Leven (2002) reported that tornado survivors in Hoisington, Kansas, were very satisfied with the donations provided by the Salvation Army. In contrast, public and some private sources provided much more total assistance, yet survivors were very critical of these sources (also see Paul 2011). One reason for this criticism is the perception that public distribution suffers from irregularities. Therefore, public authorities must ensure fair and equitable distribution among individuals, communities, and geographic areas. It should adopt a truly non-discriminatory policy to guard against unfair distribution of relief aid.

Determinants of Disaster Aid

This section will shed light on what drives donor nations to decide to provide disaster assistance to countries that have suffered these events.[10] In a number of studies, economists and others have identified the determinants of disaster aid. While studying the channel and type of disaster aid provided by donor countries, Raschky and Schwindt (2011) claimed that donors consider two arguments in providing disaster assistance: the number of fatalities caused by a disaster and the number of people affected (also see Fink and Redaelli 2011). This is consistent with Simon's (1997) finding that fatality rate not only influences people to donate, but also how much they are willing to donate. He further stated that the amount of national and international media coverage is directly related to the amount donated to disaster survivors both by individuals and by governments (also see Strömberg 2007). Becerra et al. (2014) added that disaster aid pledges are at their highest while media attention is at its peak. No doubt, TV coverage of natural or man-made disasters, known as CNN effect, is an important determinant of disaster assistance (Tornau 2006). However, Van Wassenhove (2006) noted the phenomenon occurred only for rapid onset disasters. He claimed that survivors of such an event are more likely to garner media attention than sufferers of ongoing chronic events.

In studying the determinants of the level of emergency assistance, Olsen et al. (2003) found that the extent of media coverage occasionally appeared as factor of allocation of aid by donors. Rather, they reported that such allocation is largely dependent on the degree of political interest, especially related to security, that Western donor governments had in a particular geographic region. They further claimed that the presence and strength of humanitarian stakeholders like NGOs and international organizations determined the volume of emergency aid from donor countries to the disaster-affected country. These three factors either individually or in conjunction act as determinants of emergency assistance for humanitarian crises (Olsen et al. 2003).

Becerra et al. (2014) suggested that the direct damage caused by large natural disasters, the magnitude of disaster in other words, may directly relate to the amount of emergency aid provided by foreign countries. Fatality rates, the percentage of people affected, and the amount of damage caused by a disaster all represent the severity of the event. All these factors, in turn, are associated with each other and with media reporting of a disaster. For example, the number of people killed in a disaster often correlates highly with the number of people affected by the event. However, as noted in Chapter 1, political and cultural affinity, as well as geopolitical interests, is generally positively related to the amount of emergency aid. On the other hand, Becerra et al. (2014) found no evidence of such a relationship although Strömberg (2007) and Fink and Redaelli (2011) reported a direct association between a well-established connection between donor and disaster-affected countries and the amount of emergency aid provided by the former. Fink and Redaelli (2011) further reported that donors were more likely to provide emergency assistance to the affected countries if those countries have fewer politically alliances than others.

While the above is particularly true for the USA (e.g., Drury et al. 2005; Fleck and Kilby 2010), others (e.g., Demirel-Pegg and Moskowitz 2009) maintain that US disaster aid is also motivated by humanitarian concerns. After analyzing 270 natural disasters, Fink and Redaelli (2011) reported that humanitarian need appears to predict whether a country will provide disaster aid. A similar finding was also reported by Nelson (2012), who studied the aid activity of 22 donor countries between 1997 and 2008. He found that certain political variables (e.g., trading partners, former colonies, and military allies) are significant determinants of whether donor countries provide emergency aid. The amount of aid is positively affected by these variables.

Using regression, Becerra et al. (2014) found that important determinants of post-disaster aid included level of development, country size, and the foreign reserves of the affected country. On average, countries with a higher gross domestic product (GDP) usually receive less disaster aid. Similarly, countries with more foreign exchange reserves receive less aid. This indicates that donors are influenced by their perception of the economic needs of the affected country. They also reported a reverse relationship between the amount of aid and the land area of the affected country. Fink and Radaelli (2011), however, claimed that donors favor small nations and countries that are geographically closer in providing disaster aid. Becerra et al. (2014) further reported that aid covers only 3% of the total estimated economic damage caused by disasters. In contrast, Yang (2008) claimed that aid accounted for about 73% of the cost of disaster damage. David (2011), however, reported that emergency aid does not increase after climatic disasters, but increases a very small amount following geologic events. Thus the research shows little agreement about what affects disaster aid and how much that aid actually helps long term.

In an as yet unpublished paper, Thomas Jamieson (n.d.) introduced the term "strategic compassion" as one determinant of US international emergency aid after natural disasters. He defined the term as a practice where the country provides emergency assistance to a hostile country affected by natural or man-made disasters hoping to exert political pressure in addition to providing humanitarian aid. After the 2008 Cyclone Nargis hit Myanmar, the USA responded with generous disaster aid, despite its strained relationship with the army ruler of Myanmar. Jamieson (n.d.) provided two reasons for pursuing such strategy. First, this strategy may pressure the recipient in two ways. If the country accepts the aid, it is interpreted as reduction of tension between the donor and recipient countries. If the affected country refuses the aid, disaster survivors will suffer more, which, in turn, will create civil unrest against the government. This domestic instability may benefit the donor countries. Second, providing disaster aid to a hostile country offers a better potential return.

Drawing data from Financial Tracking Service (FTS) from the United Nations Office for the Coordination of Humanitarian Affairs (OCHA) and the Center for Research on the Epidemiology of Disasters' Emergency Events Database (EM-DAT) and applying a series of ordinary least squares (OLS) regressions, Jamieson (n.d.) found that strategic incentives motivated the USA to provide emergency aid after natural disasters between 1999 and 2010, indicating that the USA provided more disaster aid to countries with which it did not agree in the U.N. General Assembly than countries with whom the USA shares similar policies. Jamieson (n.d.) represented the prevalence of national strategic incentives using the percentage of all U.N. General Assembly votes where the emergency aid recipient country agreed with the USA. He also reported humanitarian motives for US donations because he found a statistically significant positive effect with the independent variable of the number of people killed in each natural disaster considered in the study. However, the size of this effect was not large.

Strategic compassion suggests that an extreme event may create conditions for improved relationships as well as improved cooperation between foreign donors and disaster-stricken countries. Thus, international emergency aid may benefit both countries. Disasters provide opportunities for both hostile donors and long-standing rivals to improve their relationships. Rasler et al. (2013) maintain that disasters are necessary conditions for establishing such relationships. Relevant literature further suggests that natural disasters may be exploited to pursue cordial relationships between donors and recipients (Kelman 2012). Often, donors deliberately provide expecting to receive credit for such philanthropic work. This donation may also "encourage mutually beneficial bilateral trade" (Jamieson, n.d., 8). Like official development assistance (ODA), donors often provide disaster aid with cost-effectiveness in mind and hence are reluctant to offer emergency relief to countries with reputations of widespread corruption and rigid bureaucracy.

Below the international level, few studies focus determinants of disaster aid at the individual and household levels. Several studies (e.g., Tiehen 2001; Auten et al. 2002; Gittell and Tebaldi 2006; Bakija and Heim 2008) have examined what affects charitable contributions at the household level. These studies suggest a positive relationship between disaster donations, household income, level of education, religious affiliation, and tax breaks or tax incentives.[11] The magnitude of a disaster, represented by number of people affected or total number of deaths, was also directly associated with individual contributions to emergency relief fund for both disasters occurring in their own country or foreign countries. A study (Ishida 2013) conducted in Japan after the 2011 Tohoku Earthquake and 1995 Kobe Earthquake revealed that past experience, household income, savings, and age were positively associated with donations for the survivors of the two earthquakes. Ishida also reported that donations were a function of geographical distance between the disaster sites and the location of individual donors.

The literature on determinants of emergency aid is modest but growing with divergent results among the available studies. A review of relevant literature revealed that a combination of factors determines whether and how much emergency assistance is offered to assist with overseas natural disaster operations, whether to alleviate suffering, provide food, or create shelter for survivors. Further research must explore the complex relationships among determinants of emergency aid provision at both the country and household levels.

Notes

1. The behavioral approach existed before the 1970s. At that time, evaluating individual behavior during natural hazards used a logical positivist approach, which considered that people behave in an economically rational way (Montz et al. 2017). Although White in his one of the earliest works (White 1945) used a more inductive approach, he and his students added social components to the behavioral approach.
2. A reverse situation has also been reported by Islam and Walkerden (2015). In studying the disaster relief distribution after the Cyclone Side, which hit Bangladesh in 2007, they claim that NGOs generally favored their own borrowers of micro-credit. NGOs in Bangladesh and other developing countries provide micro-credit to the poor to alleviate their poverty. Wealthy households are not eligible to receive such credit from NGOs.
3. For more details, see Paul (2007).
4. Taka (Tk) 5000 was equivalent to $72 in 2007.
5. A US military helicopter crashed on May 12, 2015, killing the six US Marines on board. A commercial helicopter crashed on June 2, 2015, killing four relief workers, three of whom were volunteers with Doctors Without Border (*Kathmandu Post* 2015).
6. Although humanitarian organizations routinely claim that administrative costs are no more than 10% of their spending, they often spend 50% of their budget on those costs. For example, in a project implemented by the Adventist Development and Relief Agency (ADRA) in Nepal after the 2015 earthquakes, administrative costs were half of the budget. Save the Children, on the other hand, spent 30% of their budget in administrative costs on a project to help small businessmen in Nepal (Troutman 2015).
7. A type of outdoor latrine without any way to flush water. The structure, which is excavated in the soil, may be manufactured and installed so that odors and flies are minimized.
8. For more detail, see Paul et al. (2017).
9. A district is the second largest administrative unit in Nepal. The country is divided into 75 districts.
10. Also see Chapter 1.
11. Also see Chapter 2.

References

Agarwal, B. 2001. Participatory Exclusions, Community Forestry, and Gender: An Analysis for South Asia and a Conceptual Framework. *World Development* 29 (10): 1623–1648.

Aldrich, D.P. 2010. The Power of People: Social Capital's Role in Recovery from the 1995 Kobe Earthquake. *Natural Hazards* 56: 595–611.

Ali, A.I., and G. Ince. 2017. Distress Among Disaster-Affected Populations: Delay in Relief Provision. *Journal of Operational Research Society* 68 (5): 533–543.

Arora, V. 2015. Geopolitics Enters Nepal's Earthquake Relief Efforts: What Was Behind Nepal's Call This Week for Foreign Teams to Leave the Country? *Diplomat*, May 6.

Asia Foundation. 2016. *Nepal Government Distribution of Earthquake Reconstruction Cash Grants for Private Houses*. San Francisco: The Asia Foundation.

Auten, G.E., H. Sieg, and C.T. Clotfelter. 2002. Charitable Giving, Income, and Taxes: An Analysis of Panel Data. *American Economic Review* 92: 371–382.

Bakija, J., and B. Heim. 2008. How Does Charitable Giving Respond to Incentives and Income? Dynamic Panel Estimates Accounting for Predictable Changes in Taxation. NBER Working Paper 14237, National Bureau of Economic Research.

Balcik, B., B.M. Beamon, and K. Smilowitz. 2008. Last Mile Distribution in Humanitarian Relief. *Journal of Intelligent Transportation Systems* 12 (2): 51–63.

Becerra, O., E.A. Cavallo, and I. Noy. 2014. Foreign Aid in the Aftermath of Large Disasters. *Review of Development Economics* 18 (3): 445–460.

Blaikie, P., T. Cannon, I. Davis, and B. Wisner. 1994. *At Risk: Natural Hazards, People's Vulnerability, and Disasters*. New York: Routledge.

Bode, B. 2002. *In Pursuit of Power: The North West Institutional Analysis*. Dhaka: CARE Bangladesh.

Bolin, R., and L. Stanford. 1999. Constructing Vulnerability in the First World: The Northridge Earthquake in Southern California. In *The Angry Earth: Disaster in Anthropological Perspective*, ed. A. Oliver-Smith, and S.M. Hoffman, 89–112. London: Routledge.

Bradshaw, S. 2013. *Gender, Development and Disasters*. Cheltenham, UK: Edward Elgar.

BBC News. 2005. Where Did UK Tsunami Millions Go? http://alertnet.org/thefacts/reliefresources/112230013428.htm. Last Accessed 21 June 2006.
Burton, I., R.W. Kates, and G.F. White. 1993. *The Environment as Hazards*. New York: Guilford Press.
Christophe, S., and M.O. Adams. 2016. Return to a State of Nature, Compassionate Conservatism, Failed Response, and Their Impact on Race, Ethnicity, and the U.S. Economy: Hurricane Katrina Case Study. In *Cities and Disasters*, ed. D.C. Downey, 105–125. Boca Raton: CRC Press.
Cutter, S.L. 2003. The Vulnerability of Science and the Science of Vulnerability. *Annals of the Association of American Geographers* 93 (1): 1–12.
Cutter, S.L., and C. Finch. 2008. Temporal and Spatial Changes in Social Vulnerability to Natural Hazards. *Proceedings of the National Academy of Sciences* 105 (7): 2301–2306.
Dalit NGO Federation (DNF). 2015. *Waiting for "Justice in Response": Report of Initial Findings from Immediate Needs Assessment and Monitoring Responses Towards Affected Dalit Communities in Nepal Earthquake 2015*. DNF: Kathmandu.
Daniel, F.J., and K. Mahr. 2015. Nepal, Aid Agencies Trade Blame as Confusion Mars Quake Relief. htttp://in.reuters.com/article/earthquake-nepal-aid-idINKBNONQ1DG20150505. Accessed 2 Apr 2016.
David, A. 2011. How Do International Financial Flows to Developing Countries Respond to Natural Disasters? *Global Economy Journal* 11 (4): 1–36.
Davis, A.C., D. Lamothe, and E. O'Keefe. 2017. U.S. Response in Puerto Rico Pales Next to Actions After Haiti Quake. *The Washington Post*, Sept 28.
Delaney, P., and E. Shrader. 2000. Gender and Post-disaster Reconstruction: The Case of Hurricane Mitch in Honduras and Nicaragua. Decision Review Draft Presented to the World Bank.
Demirel-Pegg, T., and J. Moskowitz. 2009. US Aid Allocation: The Nexus of Human Rights, Democracy, and Development. *Journal of Peace Research* 46 (2): 181–198.
Donnan, S. 2005. Frustration for Donors as Aid Piles Up on Docks. *Financial Times*, May 12.
Drury, A.C., R.S. Olson, and D.A. Van Belle. 2005. The Politics of Humanitarian Aid: U.S. Foreign Disaster Assistance, 1964–1995. *Journal of Politics* 67 (2): 454–473.
Elliot, J., and L. Sullivan. 2015. How the Red Cross Raised Half a Billion Dollars for Haiti and Built Six Homes: Even as the Group Has Publicly Celebrated its Works, Insider Accounts Detail a String of Failure. June 3,

2015. https://www.propublica.org/article/how-the-red-cross-raised-half-a-billion-dollars-for-haiti-and-built-6-homes. Last Accessed 1 Nov 2017.

Emel, J., and R. Peet. 1989. Resource Management and Natural Hazards. In *New Models in Geography*, ed. R. Peet and N. Thrift, 49–76. London: Unwin Hyman.

Fink, G., and S. Redaelli. 2011. Determinants of International Emergency Aid: Humanitarian Need Only? *Word Development* 39: 741–757.

Fisher, S. 2005. Gender Based Violence in Sri Lanks in the After-Math of the 2004 Tsunami Crisis: The Role of International Organizations and International NGOs in Prevention and Response to Gender Based Violence. Phd dissertation, University of Leeds, Leeds, UK.

Fleck, R.K., and C. Kilby. 2010. Changing Aid Regimes? U.S. Foreign Aid from the Cold War to the War on Terror. *Journal of Development Economics* 91 (2): 185–197.

Fugate, W. 2012. *The State of FEMA—Leaning Forward: Go Big, Go Early, Go Fast, Be Smart*. A Report by the US Department of Homeland Security. Washington, DC.

Gittell, R., and E. Tebaldi. 2006. Charitable Giving: Factors Influencing Giving in US States. *Nonprofit and Voluntary Sector Quarterly* 35 (4): 721–736.

Government of Bangladesh (GoB). 2008. *Cyclone Sidr in Bangladesh: Damage, Loss, and Need Assessment for Disaster Recovery and Reconstruction*. Dhaka.

Hebblethwaite, B. 2011. Haiti Earthquake (2010). In *Encyclopedia of Disaster Relief*, ed. K.B. Penuel and M. Statler, 258–262. Newbury Park: Sage.

Hewitt, K. 1983. The Idea of Calamity in a Technocratic Age. In *Interpretations of Calamity from the Viewpoint of Human Ecology*, ed. K. Hewitt, 3–32. Boston: Allen & Unwin.

Hewitt, K. 1997. *Regions of Risk: A Geographical Introduction to Disasters*. London: Longman.

Holguin-Veras, J., and M. Jaller. 2012. Immediate Resource Requirements After Hurricane Katrina. *Natural Hazards Review* 15 (1): 1–12.

Holguin-Veras, J., N. Perez, S. Ukkusuri, T. Wachtendorf, and B. Brown. 2007. Emergency Logistics Issues Impacting the Response to Hurricane Katrina: A Synthesis and Preliminary Suggestions for Improvement. *Transportation Research Board* (2022): 76–82.

Hossain, S.M.M., and P. Kolsteren. 2003. The 1998 Flood in Bangladesh: Is Different Targeting Needed During Emergencies and Recovery to Tackle Malnutrition? *Disasters* 27 (2): 172–184.

Human Rights Watch. 2005. India: End Caste Bias in Tsunami Relief. https://www.hrw.org/news/2005/01/14/india-end-caste-bias-tsunami-relief. Last Accessed 16 Sept 2017.

International Agencies. 2009. *In-depth Recovery Needs Assessment of Cyclone Aila Affected Areas*. Dhaka.

International Dalit Solidarity Network (IDSN). 2013. *Equality in Aid: Addressing Caste Discrimination in Humanitarian Response*. Copenhagen: IDSN.

International Federation of Red Cross and Red Crescent Societies (IFRC). 2006. Operation Update: Pakistan Earthquake, May 30. http://www.ifrc.org/docs/appeals/05/05EA02225.pdf. Last Accessed 27 Sept 2017.

Ishida, R. 2013. Determinants of Cheritable Giving to Unexpected Natural Disasters: Evidence from Two Big Earthquakes in Japan. Unpublished Paper. http://cis.ier.hit-u.ac.jp/Japanese/society/140110ws/ishidapaper.pdf. Last Accessed 13 Sept 2017.

Islam, R., and G. Walkerden. 2015. How Do Links Between Households and NGOs Promote Disaster Resilience and Recovery? A Case Study of Linking Social Networks on the Bangladesh Coast. *Natural Hazards* 78: 1707–1727.

Jamieson, T. n.d. Strategic Compassion: The Determinants of U.S. International Emergency Aid After Natural Disasters. An Unpublished paper, University of Southern California, Los Angeles.

Jones, E.C., and A.D. Murphy (eds.). 2009. *The Political Economy of Hazards and Disasters*. Berkeley: AltaMira Press.

Kathmandu Post. 2015. Four Dead in Sindhupalchock Chopper Crash, June 2.

Kelman, I. 2012. *Disaster Diplomacy: How Disaster Affect Peace and Conflict*. Abindon: Routledge.

Kirsch, T.D., E. Leidman, W. Weiss, and S. Doocy. 2012. The Impact of the Earthquake and Humanitarian Assistance on Household Economies and Livelihoods of Earthquake-Affected Populations in Haiti. *American Journal of Disaster Medicine* 7 (2): 85–94.

Kumar, R., and T. Kala. 2006. Two Years After the Asian Tsunami: Thousands Still Suffering in India. http://www.wsws.org/articles/2006/dec2006/indi-d29shtml. Last Accessed 14 Feb 2007.

Lewis, D., and A. Hossain. 2008. A Tale of Three Villages: Power, Difference and Local Difference in Rural Bangladesh. *Journal of South Asian Development* 3 (1): 33–51.

Mahmud, T., and M. Prowse. 2012. Corruption in Cyclone Preparedness and Relief Efforts in Coastal Bangladesh: Lessons for Climate Adaptation? *Global Environmental Change* 22: 933–943.

Mallick, B., S.M. Witte, R. Sarker, A.S. Mahboob, and J. Vogt. 2009. Local Adaptation Strategies of a Coastal Community During Cyclone Sidr and Their Vulnerability Analysis for Sustainable Disaster Mitigation Planning in Bangladesh. *Journal of Bangladesh Institute of Planners* 2: 158–168.

Margesson, R. 2005. *Indian Ocean Earthquake and Tsunami: Humanitarian Assistance and Relief Operations*. Washington, DC: Congressional Research Service.

Ministry of Food and Disaster Management (MoFDM). 2009. *Special Situation Report by WASH Cluster Working Group to Disaster Management Bureau*. Dhaka: Government of Bangladesh.

Mitchell, J.T., and S.L. Cutter. 1997. *Global Change and Environmental Hazards: Is the World Becoming More Disastrous?* Washington, DC: AAG.

Montz, B.E., G.A. Tobin, and R.R. Hagelman III. 2017. *Natural Hazards: Explanation and Integration*. New York: The Guilford Press.

Mustafa, D. 2003. Reinforcing Vulnerability? Disaster Relief, Recovery, and Response to the 2001 Flood in Rawalpindi, Pakistan. *Environmental Hazards* 5: 71–82.

Nadiruzzaman, M., and D. Wrathall. 2014. Participatory Exclusion—Elite Capture of Participatory Approaches in the Aftermath of Cyclone Sidr. UNU-EHS Working Paper Series 3, United Nations University, Bonn.

Nadiruzzaman, M., and D. Wrathall. 2015. Participatory Exclusion—Cyclone Sidr and Its Aftermath. *Geoforum* 64: 196–204.

Nagurney, A., A.H. Masoumi, and M. Yu. 2015. An Integrated Disaster Relief Supply Chain Network Model with Time Targets and Demand Uncertainty. In *Regional Science Matters: Studies Dedicated to Walter Isard*, ed. P. Nijkamp, A. Rose, and K. Kourtit, 287–318. New York: Springer.

Nagurney, A., and L.S. Nagurney. 2016. A Mean-Variance Disaster Relief Supply Chain Network Model for Risk Reduction with Stochastic Link Costs, Time Targets, and Demand Uncertainty. In *Dynamics of Disasters: Key Concepts, Models, Algorithms, and Insights*, ed. L.S. Kotsireas, A. Nagurney, and P.M. Pardalos, 231–255. New York: Springer.

Nelson, T. 2012. Determinants of Disaster Aid: Donor Interest or Recipient Need? *Global Change, Peace & Security* 24 (1): 109–126.

Nissanka, J. 2006. Tsunami Recovery Process: Reorganization on Way. http://www.sundayobserver.lk.2005/12/25/news01.html. Last Accessed 19 Aug 2011.

Olsen, G.R., N. Carstensen, and K. Heyen. 2003. Humanitarian Crises: What Determine the Level of Emergency Assistance? Medical Coverage, Donor Interests and the Aid Businees. *Disasters* 27 (2): 109–126.

Pariyar, R. 2015. Double Trouble: The Dalit Community Has Faced Discrimination in Relief Distribution After the Great Quake. *Kathmandu Post*, Dec 6.

Paul, A., and M. Rahman. 2006. Cyclone Mitigation Perspectives in the Islands of Bangladesh: A Case of Sandwip and Hatia Islands. *Coastal Management* 34: 199–215.

Paul, B.K. 1998. Coping with the 1996 Tornado in Tangail, Bangladesh: An Analysis of Field Data. *The Professional Geographer* 50 (3): 287–301.

Paul, B.K. 2003. Relief Assistance to 1998 Flood Victims: A Comparison of the Performance of the Government and NGOs. *The Geographical Journal* 169 (1): 75–89.

Paul, B.K. 2006. Disaster Relief Efforts: An Update. *Progress in Development Studies* 6 (3): 211–223.

Paul, B.K. 2007. 2004 Tsunami Relief Efforts: An Overview. *Asian Profile* 35 (5): 467–478.

Paul, B.K. 2011. *Environmental Hazards and Disasters: Contexts, Perspectives and Management*. Hoboken, NJ: Wiley-Blackwell.

Paul, B.K., and J. Leven. 2002. *Emergency Support Satisfaction Among 2001 Hoisington, Kansas Tornado Victims*. Quick Response Report #154. NHRAIC, University of Colorado, Boulder, CO.

Paul, B.K., and M.K. Rahman. 2013. Recovery Efforts: The Case of the 2007 Cyclone Sidr in Bangladesh. In *Natural Disasters and Adaptation to Climate Change*, ed. S. Boulter, J. Palutikof, D.J. Karoly, and D. Guitart, 167–173. New York: Cambridge University Press.

Paul, B.K., B. Acharya, and K. Ghimire. 2017. Effectiveness of Earthquakes Relief Efforts in Nepal: Opinions of the Survivors. *Natural Hazards* 85: 1169–1188.

Paul, S.K., and M.N. Hossain. 2013. People's Perception About Flood Disaster Management in Bangladesh: A Case Study on the Chalan Beel Area. *Stamford Journal of Environment and Human Habitat* 2: 72–86.

Paul, S.K., B.K. Paul, and J.K. Routray. 2012. Post-Cyclone Sidr Nutritional Status of Women and Children in Coastal Bangladesh: An Empirical Study. *Natural Hazards* 64 (1): 19–36.

Pelling, M. 1998. Participation, Social Capital and Vulnerability to Urban Flooding in Guyana. *Journal of International Development* 10: 469–486.

Pelling, M. 1999. The Political Ecology of Flood Hazard in Urban Guyana. *Geoforum* 30: 249–261.

P'Rayan, A. 2005. Impact of Tsunami Relief and Rehabilitation Work in India. http://www.Worldpress.org/Asia2116.cfm. Last Accessed 13 June 2005.

Raschky, P.A., and M. Schwindt. 2011. On the Channel and Type of Aid: The Case of International Disaster Assistance. *European Journal of Political Economy* 28 (1): 119–131.

Rasler, K., W.R. Thompson, and S. Ganguly. 2013. *How Rivalries End*. Philadelphia: University of Pennsylvania Press.

Reliefweb. 2005. Indonesia: Aceh Tsunami Aid Sold Outside the Tsunami Ravage Province. http://www.reliefweb.int/rw.RWB.FSF/dh900SID?EVOD-6J5DUK?OpenDocument. Last Accessed 4 Aug 2010.

Reliefweb. 2013. Equality in Aid: Addressing Caste Discrimination in Humanitarian Response, 30 Sept, 2015. https://reliefweb.int/report/world/equality-aid-addressing-caste-discrimination-humanitarian-response. Last Accessed 6 Nov 2017.

Robbins, P. 2012. *Political Ecology*. Boston: Wiley.

Sanderson, D., and B. Ramalingam. 2015. *Nepal Earthquake Response: Lessons for Operational Agencies*. ALNAP Lessons Paper. London: ALNAP/ODI.

Simon, A. 1997. Television News and International Earthquake Relief. *Journal of Communication* 47 (3): 82–93.

Specia, M. 2015. Nepal's Poorest Getting the Least Earthquake Aid, Activists Warn. http://mashable.com/2015/06/02nepal-caste-earthquake-relief/#pW-wbQKJa5kqy. Last Accessed 24 Feb 2016.

Strömberg, D. 2007. Natural Disasters, Economic Development, and Humanitarian Aid. *Journal of Economic Perspectives* 21 (3): 199–222.

Susila Dharma International (SDI). 2015. The Caste System and Disaster Relief in Nepal, 8 June. https://www.susiladharma.org/the-caste-system-and-disaster-relief-in-nepal/. Last Accessed 6 Nov 2017.

Susman, P., P. O'Keefe, and B. Wisner. 1983. Global Disasters: A Radical Interpretation. In *Interpretations of Calamities*, ed. K. Hewitt, 274–276. London: Allen & Unwin.

Tiehen, L. 2001. Tax Policy and Charitable Contributions of Money. *National Tax Journal* 54 (4): 707–823.

Tornau, U. 2006. CNN Effect: Power or Mean? A Study of Media Influence on Foreign Policy Decision—Making Comparative Analysis of Two Humanitarian Disasters: Ethiopia (1984–1985) and Somalia (1992).

Master of Comparative History, Utrecht University, Utrecht, The Netherlands.
Transparency International Bangladesh (TIB). 2011. *Annual Report 2010.* Dhaka: TIB.
Troutman, E. 2015. What Happened to the Aid? Nepal Earthquake Response Echoes Haiti. *Aid Works,* 19 June. http://aid.works/2015/06/nepal-haiti/. Last Accessed 16 Sept 2017.
U.N. (United Nations). 2006. *Tsunami Recovery: Taking Stock After 12 Months.* New York.
Van Wassenhove, L. 2006. Humanitarian and Logistics: Supply Chain Management in High Gear. *Journal of the Operational Research Society* 57: 475–489.
Versluis, A. 2014. Formal and Informal Material Aid Following the 2010 Haiti Earthquake as Reported by Camp Dwellers. *Disasters* 39 (S1): S94–S109.
Walton-Ellery, S. 2009. *A Review of the Cyclone Aila Response 2009: IFRC-Led Emergency Shelter Coordination Group*. Dhaka: IFRC.
Walton, R., R. Mays, and M. Haselkorn. 2011. Defining "Fast": Factors Affecting the Experience of Speed in Humanitarian Logistics. In *Proceedings of the 8th International ISCRAM Conference*, Liston, May.
Washington Post. 2006. Corruption, Fraud Hunt Tsunami Efforts, Sept 25.
Weesjes, E. 2015. From the Editor. *Natural Hazards Observer* 39 (5): 3.
Wescoat J.L., Jr. 1987. The Practical Range of Choice in Water Resources Geography. *Progress in Human Geography* 11 (2): 41–59.
White, F. 1945. Human Adjustment to Flood: A Geographical Approach to the Flood Problem in the United States. Doctoral Dissertation, Research Paper No. 29, University of Chicago Press, Chicago.
Wisner, B., P. Blaikie, T. Cannon, and J. Davis. 2004. *At Risk: Natural Hazards, People's Vulnerability and Disasters*. London: Routledge.
Yang, D. 2008. Coping with Disaster: The Impact of Hurricanes on International Financial Flows. *B.E. Journal of Economic Analysis & Policy* 8 (1): 1–45.
Zakaria, R. 2015. The Deceptions of Disaster Relief in Nepal: The Aid-Industrial Complex is in Dire Need of Overhaul. *Aljazeera America*, July 2. http://america.aljazeera.com/opinions/2015/7/the-deceptions-of-disaster-relief-in-nepal.html. Last Accessed 16 Sept 2017.

4

Response to and Emergency Relief Efforts for the Selected Disasters

This chapter provides an overview of the nature and extent of emergency relief efforts undertaken in countries impacted by the 2004 Indian Ocean Tsunami (IOT), 2005 Hurricane Katrina, 2005 Pakistan/Kashmir and 2010 Haiti Earthquakes, and current famine in South Sudan, along with essential basic statistics regarding these disasters. The chapter reviews shortcomings, successes, challenges, and other pertinent features of each one of these disaster relief operations. This review is essential for understanding disaster response and guiding efficient, orderly, and sustainable relief operation plans for survivors of future natural disasters. Further, the chapter provides valuable lessons for all participant agencies, individuals, and particularly decision makers and disaster managers regarding what should or should not be done during the relief phase of the disaster cycle.

Depending on the magnitude and type of the event, disaster relief operations generally last for a few weeks to almost a year. Inadequate and/or delayed operations create a human-induced tragedy that exacerbates and prolongs the plight of those already suffering the effects of a natural disaster. Such response also reflects failure of government agencies, public administrators, and others to provide adequate emergency

B. K. Paul, *Disaster Relief Aid*,
https://doi.org/10.1007/978-3-319-77282-0_4

support to disaster survivors and sufferers (Paul 2011). To demonstrate disaster relief operations, this chapter reviews four such disasters (rapid onset) and one complex emergency (slow onset) of the more than 100 major natural disasters and complex emergencies that have occurred globally since the beginning of this century.

The selected extreme events cover over 13 years, four continents (Africa, Asia, North America, and Latin America), and illustrate a range of relief efforts. They include three types of natural disasters (tsunami, hurricane/cyclone/typhoon, and two earthquakes) and one complex emergency (South Sudan Famine). Three (2004 Indian Ocean Tsunami, 2010 Haiti Earthquake, and 2005 Kashmir Earthquake) of the four natural disasters in this study were among the five deadliest since 2000 (Real World Survivor 2015). Although each extreme event is unique, they all share many commonalities among their relief operations. Thus, this review will provide broader understanding of the attributes of such operations.

The Indian Ocean Tsunami (IOT)

On December 26, 2004, a 9.0-magnitude earthquake occurred in the ocean off the west coast of northern Sumatra, Indonesia, 18.6 miles (30 km) below the sea level (Rofi et al. 2006).[1] It happened where the Indian plate slipped underneath the Burma plate. This pushed the top plate up as much as 15 feet (4.57 meters) over 600 miles (965 km) of its length. The result was a ripple of major tidal waves, which crashed onto the Sumatra coast 15 minutes after the quake with waves as high as 100 feet (30.48 meters). These waves hit Thailand 75 minutes after the earthquake and hit Sri Lanka and India four hours after the quake. Waves even reached as far as Somalia and Kenya, some 3100 miles (5000 km) away from the epicenter. It was the most deadly tsunami in recorded history (U.N. 2006).

This tragic event impacted as many as 14 countries in Asia and Africa.[2] However, sections of Indonesia, Sri Lanka, India, and Thailand have suffered the worst devastation. The tsunami killed over 200,000 from over 50 countries and damaged homes of over one million people in 14 countries bordering the India Ocean (Karan 2011).[3] The four most affected countries, India (16,269), Indonesia (167,540), Sri Lanka (35,322), and Thailand (8212), accounted for 99.75% of all deaths caused by this extreme natural event (Karan 2011). In particular, the Aceh was the worst affected region of Indonesia because of its proximity to the epicenter of the earthquake. This province was also a conflict zone between the Indonesian government and the pro-independence Free Aceh Movement for 29 years.

In addition to the large number of local residents, at least 9000 foreign tourists (mostly European) are among the dead or missing. The European nation that suffered the most casualties was Sweden, whose death toll was 428, with 116 missing. Other foreign countries whose citizens died and/or were injured include Australia, Britain, France, Germany, Japan, Nigeria, Norway, South Africa, South Korea, Tanzania, and the USA. Moreover, more than 3000 miles (4500 km) of road and 118,000 fishing boats were damaged or destroyed, while about 5 million people needed some form of emergency relief: medical attention, shelter, clothes, and food (U.N. 2006).

Measured in terms of lives lost, the 2004 Indian Ocean Tsunami is the single worst tsunami in history. It was caused by the second largest earthquake ever recorded on a seismograph and the worst natural disaster the United Nations and the global community had faced in its 60 years of operation. This earthquake-induced tsunami caused an enormous environmental impact that will affect the region for many years to come. The tsunami not only destroyed governments' ability to respond, but also adversely influenced the livelihoods of people residing in the coastal communities of affected countries. In monetary terms, the Indian Ocean Tsunami of 2004 caused $10 billion in damage in a day (U.N. 2006).

Global Generosity

The international community's response to the 2004 tsunami was immediate and overwhelming. In several impacted areas, relief work began immediately following the disaster, with domestic and international emergency teams commencing work within hours. Five United Nations Disaster Assessment and Coordination teams were quickly deployed to five tsunami-affected countries of Asia. Thirty-four countries, including Australia, Germany, India, Japan, New Zealand, Singapore, the UK, and the USA, dispatched military personnel and equipment, and public health specialists to the tsunami-devastated countries to distribute aid, remove corpses, clear rubble, provide immediate medical attention, and offer psychological counseling (Paul 2007). The Singapore government also offered the use of its air and naval facility as a base for U.N. operations and other relief agencies, as well as to countries such as Australia, the USA, France, and Japan who were undertaking relief efforts.

At the beginning of 2005, the United Nations launched a Flash Appeal in the amount of $1.1 billion for immediate needs, 85% of which was pledged within four weeks (Valley 2006).[4] By early February 2005, governments from more than 40 countries, U.N. agencies, non-governmental organizations (NGOs) of developed countries, international financial institutions, as well as private and public entities around the world had pledged $5.5 billion in emergency assistance. Based on the Financial Tracking Service (FTS) of the United Nations Office for the Coordination of Humanitarian Affairs (UNOCHA), as of 31 December, 2016, the total pledged amount by all sources was $6832 million, of which $6256 million or 92% has already been released (Table 4.1). Sixty-two percent of these funds came from private sources, and the rest came from official government sources (Table 4.1).

Large (2005) reported that most funds provided by individuals have gone to NGOs participating in the relief efforts and that most international NGOs (INGOs) were adequately funded by private sources. Meanwhile, nearly 70% of private donations went to just 12 agencies: The Red Cross movement, UNICEF, and the 10 largest INGOs or INGO consortia (Flint and Goyder 2006). Consequently, much official funding went to U.N. organizations. However, several official funders bypassed U.N. organizations and prepared their own assessment primarily because of their lack of confidence in U.N. assessments (de Ville de Goyet and Moriniere 2006).

Table 4.1 Total funding per donor (appeal and other[a]) as of December 31, 2016, (in US$ million) for the 2004 IOT

1	2	3	4	5
Donor	Fund provided (% of grant total)	Outstanding pledges	Total (2+3)	Fund provided as % of total [(2/4)100]
Private (individuals and organizations)	3892 (62)[b]	53	3945	99
Allocation of funds from Red Cross/Red Crescent	256 (4)	0	256	100
European Commission	173 (3)	45	218	79
Allocation of unearmarked funds by U.N. agencies	24 (<1)	0	24	100
Allocation of unearmarked funds by IGOs	200 (3)	0	200	100
Islamic Development Bank	150 (2)	0	150	100
Governments of non-affected countries	1561 (25)	478	2039	77
Total	6256 (100)	576	6832	92

[a]Includes contributions to the appeal and additional contributions outside the appeal (e.g., bilateral and Red Cross)
[b]All percentages are rounded to the nearest whole number
Source Prepared by the author based on http://fts.unocha.org (Table B), December 31, 2016

The financial response to this disaster broke all records with pledges of $13.6 billion for relief and recovery works (Flint and Goyder 2006). This included an unprecedented response from individuals and firms of $4 billion toward the relief effort. UNICEF received almost twice as much as it had sought in donations. The World Health Organization (WHO) and the World Food Program (WFP) secured 95% of their funding objectives. As many as 99 countries (77 of which are not members of the Development Assistance Committee (DAC) of the Organization for Economic Co-operation and Development (OECD)) and three intergovernmental organizations (the European Commission, the Organization of Petroleum Exporting Countries (OPEC), and the African Union) contributed to this response. Funding from private sources was also generous, fast, and flexible. With more than $20.5 million, online donations set a record in the UK (Telford and Cosgrave 2007).

The 2004 IOT aid amounts given (in US$ million) by the top 10 donors as of December 31, 2016, are presented in Table 4.2. The table shows that private donations accounted for slightly over 62% of all aid. This indicates that response by individuals living in non-affected countries, particularly in Western countries, was dramatic and spontaneous. For example, nearly every adult Briton donated to the tsunami victims in some way. In the USA, the tsunami sparked the greatest outpouring of donations the country has ever seen in response to a foreign disaster (Valley 2006). Clark (2005, 384) wrote, "ordinary people's giving exceeded or pre-empted that of their governments: they pledged money, they gave clothes and other necessities, and many offered their time

Table 4.2 Amount of disaster aid (in US$ million) funded to tsunami-affected countries by the top 10 donors as of December 31, 2016

Donor	Funding	% of Grand total
Private (individuals and organizations)	3892	62.2
Japan	502	8.0
Allocation of funds from Red Cross/Red Crescent	256	4.1
European Commission	173	2.8
UK	137	2.2
USA	134	2.1
Germany	126	2.0
Canada	117	1.9
Norway	86	1.4
China	62	1.0
Total	5485[a]	100.0

[a]This means other sources provided $771 million, which is 12% of the of the grand total

Source Prepared by the author based on Financial Tracking Service (http://fts.unocha.org—last accessed 23 March 2017)

or skills." Ultimately, the global response to the 2004 Indian Ocean Tsunami was the largest on record.

Data presented in Table 4.2 show that Japan was the single biggest donor to the tsunami response, providing $502 million and deploying more than 100 emergency workers to assist with relief efforts (Kweifio-Okai 2014). By donating $256 million, Red Cross/Red Crescent ranks second, followed by the European Commission which contributed $173 million. The remaining donors in descending order of contribution are as follows: The UK, USA, Germany, Canada, Norway, and China (Table 4.2). All sources combined provided $6.3 billion, which is 52% of the pledged amount.[5] However, the gap between pledged and released amount is substantial. Iran, Saudi Arabia, and Russia each provided only 13% of their pledged amount, and the USA contributed 38% of the pledged amount. Six countries (Brunei, Cameroon, Georgia, Jamaica, Macedonia, and Niger) made pledges but did not contribute anything. In contrast, 64 countries, including Japan, Germany, Canada, the Netherlands, Denmark, Greece, and Switzerland, provided 100% of their pledged amount (Financial Tracking Service 2017).

According to the U.N. OCHA's Financial Tracking Service (2017), Indonesia received the most humanitarian aid with more than $1 billion. With nearly $652 million, Sri Lanka followed Indonesia as the second largest recipient of aid. Other countries that received notable aid are listed in order: India ($151 million), Maldives ($91 million), Thailand ($30 million), and Somalia ($15 million). Other three affected countries (Malaysia, Myanmar, and Seychelles) each received less than $10 million (Kweifio-Okai 2014).

Numerous private and public organizations, as well as individuals, from more than 62 (both affected and non-affected) countries have been involved in the relief operation. Two tsunami-impacted countries (India and Malaysia) also provided assistance to other affected countries. In fact, India coordinated relief operations in Indonesia, the Maldives, and Sri Lanka. India and Thailand initially refused to accept foreign aid for tsunami survivors, saying that they could cope on their own. By doing this, both countries wanted to show that they are economically and politically better-off than compared to other tsunami-ravaged countries. There is another reason for India's initial refusal to seek emergency assistance; acceptance of foreign aid meant allowing foreign forces and individuals to have access to all tsunami-affected parts of India. However, outsiders are not allowed to visit the remote Andaman Islands due to military reasons (Ugra 2005). India was even "furious" with Sri Lanka for not providing the information that US troops were arriving in Colombo.

No other disaster in the world to date has received as much direct participation from both organizations and private individuals. As many as 39 U.N. agencies, military forces from 40 countries, nearly 60 foreign corporations, and 700 NGOs of both developed and developing countries participated in the 2004 IOT relief efforts (Chia 2007). Asian expatriates around the world contributed to the relief efforts by offering aid and technical support to their tsunami-affected countries of origin, while the U.N. deployed coordination officers to key locations for logistical support (U.N.-ESC 2005). All told the combined response was unprecedented.

The 2004 Indian Ocean Tsunami prompted a substantial, heartfelt response, particularly from the West, for several reasons. First, not only were local residents affected, but several thousand Westerners on vacation in the tsunami-affected countries were also victims. This gave many Europeans and Americans a personal connection to the disaster. Second, the massive print and electronic media coverage prompted an extraordinary global response. Some may have found a link between this coverage and Western victims in that the disaster occurred on Boxing Day, when the Western world could be made to feel guilty about its festive overindulgence. This tsunami also came early in the budget year for most governments, which may account for the unprecedented response.

Finally, the tsunami occurred in a region of political and economic interest to governments of developed countries and to large private companies (Spiegel 2005). For instance, many in the international community view Washington, DC's involvement in the Aceh Province of Indonesia as a way of trying to bolster its image in the Muslim world (Brown 2005). The 2004 tsunami provided an opportunity for the US government to demonstrate that its "war on terrorism" is not to be confused with a war on Islam. Such unbiased support by the USA succeeded largely in a softening of its image in several Muslim countries.

Local Generosity

Generosity among local people in the tsunami-afflicted areas was equally impressive. Clark (2005) provided an example of such generosity from the weblog of an Indian journalist, Amit Varma, who traveled through the state of Tamil Nadu, India, following the tsunami. On January 3, 2005, he posted the following message:

> December 26 could have been the happiest day of Rafiq's life if the tsunami hadn't struck – he was supposed to get married [in] that day. His *nikah* was fixed for noon, but the waves came in while it was still morning, and the marriage was cancelled. Rafiq was in the village of Parangipettai, close to a number of affected villages. Instantly, all the men of the community mobilised themselves under the Jamaat, their local organization, and swung into action.
>
> They took all the veg biryani that had been prepared for the wedding feast, and went and fed it to the affected people. From that day until the day we met them, a week after the tsunami, they fed breakfast and lunch to the affected people, making either lemon rice or veg biryani. They mobilised their funds superbly, and were well networked through mobile phones. If any village ran short of food, one phone call was all it would take to bring volunteer[s] rushing over with more food.
>
> Interestingly, even after the government set up its own operation, a few days later, the local people still requested the Jamaat to keep feeding them, and the Jamaat agreed. A deep bond had been formed between the villagers, who were all Hindus, and these Muslim men who rushed to help their neighbors because they believed that to be the way of their religion. (Clark 2005, 385–386)

Local people of tsunami-impacted countries not only provided food to the disaster survivors, but they also provided shelter and helped to search for and locate families and friends of the victims, even before they sought out their own relatives in the affected areas. This spontaneity of local solidarity in the immediate aftermath of the tsunami demonstrates the potential of the affected countries to take charge after extreme natural events and dispels the notion that they are passive victims.

Competition and Coordination

Almost all countries that initially pledged donations to the 2004 Indian Ocean Tsunami have increased their pledged amount over time. For example, the USA originally promised only $15 million for this disaster, but in response to the criticism of U.N. Under-Secretary-General for Humanitarian Affairs, Jan Egeland, the US government added another $20 million to its original pledge on December 28, 2004, after Egeland categorized charitable contributions of rich countries as "stingy" (Margesson 2005). As the death toll rose and contributions from Europe soared, the USA again increased its pledge to $857 million (Wehrfritz and Cochrane 2005). Additionally, President Bush asked his father, former President George H. W. Bush, and former President Bill Clinton to lead a US effort to provide private aid to the tsunami victims.

During the first few days after the tsunami, major donors were engaged in a veritable race to show who was most generous. Japan, for example, wanted to exceed the amounts offered by China and the USA to show politically that Japan is economically more powerful than China. As an Asian country, Japan saw itself as needing to shoulder more of the relief burden than the USA since the disaster occurred in Asia (Tamamoto 2005).[6]

Many commentators claim excessive and competitive donor responses to the 2004 Indian Ocean Tsunami threaten less dramatic but equally important relief efforts elsewhere. For example, British Prime Minister Tony Blair expressed concern that tsunami aid could detract from other pressing development needs. Blair pointed out that there was a disaster comparable to a "preventable tsunami every week in Africa," where 10,000 people die daily from AIDS and malaria alone. In the same context, experts refer to the current crisis in Darfur, Sudan, and maintain that Darfur deserves the same kind of attention and response as the 2004 tsunami received. The civil war between the Sudanese government and Darfur guerrillas has resulted in massive genocide. The number killed and displaced in Darfur probably more than 2.5 million has been astounding for a war that began only two years ago (Kasfir 2005).[7]

Not withstanding these criticisms, donor governments have provided aid to tsunami survivors via United Nations agencies, NGOs, and as bilateral aid directly to national governments of impacted countries. However, considering the scale of the 2004 tsunami disaster and concerns over proper relief distribution, the U.N. stepped in and coordinated all international relief efforts to possibly eliminate, or at least reduce, duplicate supply chains. Initially, the USA, with help from Australia, India, and Japan set up a "core-group" immediately after the disaster to assure aid reached targeted populations. The U.N. subsequently took over this work and ran what has become the world's largest humanitarian relief operation in history.

Despite many flaws, the emergency relief efforts of this well-publicized disaster met its short-term objectives of saving many lives and meeting the emergency needs of the affected population. Clearly, rapid mobilization of relief goods met the urgent needs of tsunami survivors. While corruption, coordination problems, and biases occurred in disbursing emergency relief, still virtually all of the tsunami survivors received food, shelter, and medical assistance.[8] "As a result of the immediate relief response, food crises, outbreaks of diseases, and other predicted negative outcomes were avoided. Moreover, along with the unprecedented global private and public response, both national and international military personnel were instrumental in rendering aid, especially among hard-to-reach populations" (Paul 2007, 476).

Hurricane Katrina

Katrina was the deadliest hurricane since 1900, and it was the costliest natural disaster on record in the USA, causing $108 billion in damage (Zimmermann 2015). This hurricane made landfall along the low-lying delta areas of the southeast Louisiana coast as a Category 3 storm on Sunday, August 28, 2005. Increasing in intensity, in the early morning hours of Monday, August 29, the eye of Hurricane Katrina passed over the city of New Orleans. Katrina was the eleventh named storm, the fifth hurricane, and only the second Category 5 hurricane of the 2005 Atlantic hurricane season. In addition, it was the sixth strongest Atlantic hurricane ever recorded. Katrina formed over the Bahamas on August 23, 2005, and crossed southern Florida as a Category 1 hurricane before strengthening rapidly in the Gulf of Mexico (HRC 2006).

This hurricane was associated with a storm surge of waters 15–20 feet (4.6–6.1 meters) high along the coastlines of Louisiana, Mississippi, and Alabama. Katrina damaged more than a million housing units, causing the displacement of approximately one million Gulf Coast residents to communities in every state from Maine to Hawaii. Nearly half of the displaced people were from the city of New Orleans. This city withstood the pummeling storm surge and battering hurricane winds, but succumbed to a technological catastrophe because of post-storm failure of the human-engineered levee system. This produced massive flooding, which was to become the major hazard (Zimmermann 2015).

Moreover, the majority of those displaced by Katrina were African–Americans. This hurricane caused the largest displacement of Americans since the civil war. An estimated 1833 people died in the hurricane, and an overwhelming majority of them were from Louisiana. Katrina battered offshore oil platforms and pipelines as well, shutting down oil production in much of the Gulf (Elliot and Pais 2006). The hurricane also affected the casino and entertainment industry, as many of the Gulf Coast's casinos were either destroyed or sustained considerable damage. The total reduction in oil production from the Gulf of Mexico in the six-month period following Hurricane Katrina was approximately one-fourth of annual production, and the deficit in gas production over the same period was about one-fifth (Cruz and Krausmann 2008).

Although the Katrina crisis was precipitated by a physical event, it was the failure of social and political systems that turned this extreme event into a human catastrophe. Katrina's impact is not just economic or material. It has affected the political, educational, and social infrastructure of New Orleans, leaving the once-vibrant city in dire need of assistance. It was clear that poverty and race played substantial roles before, during, and in the aftermath of the disaster, and this circumstance re-emphasizes the socioeconomic and racial inequalities in American society (Elliott and Pais 2006; Lavelle and Feagin 2006; Elder et al. 2007).

Although Hurricane Katrina affected the coastal areas of three Gulf Coast States, the people of New Orleans suffered disproportionately from this disaster. Flooding caused by levee failures in the wake of the hurricane added to the misery. Even before the storm reached New Orleans, three to eight feet (0.9–2.4 meters) of water flooded part of it and eventually inundated more than 80% of the city (Jonkman et al. 2009). Heavy winds and storm surges caused by Katrina severely weakened the city's levee system. Because of widespread flooding, the public authorities initiated a massive search and rescue (SAR) operation in New Orleans. This hurricane left an estimated three million people in the affected areas without electricity for an extended period of time.

Official Response: Domestic

There is no doubt that officials at all levels, from the White House, to statehouses to city halls, failed miserably to properly respond to Hurricane Katrina. Newspapers in the USA and abroad have accused the US government for failing to take the crisis seriously enough to avoid preventable chaos and suffering. A report prepared by an 11-member all Republican, Congressional panel also acknowledges leadership failures before, during, and after Katrina struck the Gulf Coast. In the words of Jeffrey Jenson (2005, 195), "Federal, state, and local government responses in the initial hours and days following

Katrina were characterized by disorder, chaos, and ineffective coordination." Michael Chertoff, Secretary of the Department of Homeland Security (DHS), himself acknowledges that the response to Katrina was unacceptable and inadequate at all levels (Stout 2005).

Based on its investigation into the disaster, the US Senate's Committee on Homeland Security and Governmental Affairs released a special report titled *Hurricane Katrina: A Nation Still Unprepared* (749 pp) in 2006 that addressed the purpose of the investigation: to better understand how and why failures occurred and to make recommendations to strengthen US emergency preparedness and response efforts. The Committee identified four factors that contributed to the response failures: Long-term warnings went unheeded, and government officials neglected their duties to prepare for a forewarned catastrophe; government officials took inefficient actions or made poor decisions in the days immediately before and after landfall; systems on which officials relied to support their response efforts failed; and government officials at all levels failed to provide effective leadership. The Committee believed that coordination among government officials at various levels also failed.

The USA Today, in conjunction with CNN and the Gallup Poll and with the cooperation of the Red Cross, conducted a telephone survey on Hurricane Katrina from September 30 through October 9, 2005. The survey randomly selected 1510 survivors of Hurricane Katrina, aged 18 and over from nearly 470,000 families who registered with the Red Cross for assistance after the hurricane hit the Gulf Coast on August 29, 2005. These respondents were from all three affected states, and they were asked among other issues to rate federal, state, and local government response to the hurricane and the aftermath. Gallup took several steps to ensure the representativeness of its sampling and extended its field period to ten days to include up to nine contact attempts (Elliot and Pais 2006).

Table 4.3, which presents ratings of the respondents, shows that an overwhelming majority of New Orleans respondents rated performance of all three levels of government as only fair to poor. In contrast to this, majority respondents from Alabama rated performance of all three governments as excellent to good, while more than half of all respondents

Table 4.3 Opinions of respondents (in %) regarding performance of federal, state, local government on their response to Hurricane Katrina and the aftermath (September 30–October 9, 2005)

City/state	Federal government	State government	Local government
New Orleans			
Excellent/good	24	21	33
Only fair/poor	75	78	65
Louisiana			
Excellent/good	31	30	47
Only fair/poor	67	68	51
Mississippi			
Excellent/good	42	58	56
Only fair/poor	53	40	41
Alabama			
Excellent/good	56	67	72
Only fair/poor	43	30	26

Source Prepared by the author based on information provided by Jones and Carroll, October 18, 2005

of Mississippi rated state and local government response as excellent to good. The table further suggests that among all three governments, performance of local government was best, followed by that of state and federal governments.

A number of factors were associated with government failure to act with sufficient urgency as well as the inadequacy of its response to Hurricane Katrina. The Federal Emergency Management Agency (FEMA) failed to act properly and in a timely manner for several reasons. Since the formation of the DHS after the September 11, 2001, terrorist attacks, the administration virtually abandoned FEMA, leading to a mass exodus of experienced professionals from that agency. Michael Brown, FEMA director at the time of Katrina's landfall, was a former aide to President Bush, a political appointee lacking professional background for the position. Such an appointment with no emergency management experience was inherently unethical. It is now hard to imagine that FEMA had at one point been a very highly respected part of the government.

Critics claim that FEMA had been systematically neglected, underfunded, and integrated into the recently formed DHS (Stout 2005).

William Waugh Jr. (2005, 7) claims, "FEMA had been stripped of functions and disconnected from the national networks of governmental and non-governmental disaster relief organizations upon which the nation has historically depended during major disasters. Funding cuts, transfer of critical elements to other DHS units, changes in policy priorities, changes in plans and procedures and organizational structures without adequate involvement of the professional emergency management and disaster research communities and without adequate investments in training; ....and retirements of experienced FEMA personnel all contributed to the loss of emergency management capacity." However, failure of the emergency management system to respond quickly and effectively to Hurricane Katrina was not unexpected to hazard researchers and others involved with disaster management.

Critics also believe that the slow and inadequate response was associated with the diversion of funds, personnel, and equipment to Iraq and Afghanistan. Many residents of hurricane-affected areas, including New Orleans, were poor African–Americans, historically supporters of the Democratic Party. Critics argued that money needed to shore up levees in and around New Orleans was diverted to tax cuts, whereas proper maintenance of the levee system should have been addressed decades before. The US Army Corps of Engineers had allocated funding for at least some levee work, but that funding was cut by a record $71.2 million in June of 2004 (Waugh Jr. 2005). Levees for the largely black Ninth Ward were in the worst condition, while areas with higher land values were given priority. Thus, both social and political factors might have played a role in the slow reaction by emergency management entities.

There are many examples of the failures of the emergency management system to respond quickly and effectively to Hurricane Katrina. For example, state and city leaders in New Orleans were late in issuing evacuation orders (Waugh Jr. 2005). Similarly, emergency response officials in Mississippi issued mandatory evacuation when Katrina was just hours away from landfall. Interviews conducted in New Orleans reveal that about one-fifth of all respondents did not evacuate before Katrina's landfall (Travis 2005; Harrington et al. 2006); also, the rate of evacuation was nearly 20% higher among Caucasians than African–Americans (also see Elder et al. 2007).

Most of the individuals who were trapped in New Orleans could not leave because they had no money, no means of transport, and nowhere to go. Nearly 33% of New Orleans residents did not own a car, yet the official emergency plan called for people to evacuate by car. Even if they had had a car, many individuals, particularly single-mothered households, were unable to manage the logistics of leaving (including finding money for gas or hotel rooms if they did leave). This is in part because the hurricane hit at the end of the month, when paychecks were already spent. Some did try to leave, but were turned back as the roads were blocked with traffic (Butterbaugh 2005).

Plans to evacuate residents of New Orleans before the city flooded, and to rescue them after, were incomplete and inefficient. Some preferred to stay in a social environment that they trusted rather than fleeing to safer, but unfamiliar surroundings. Some stayed because they could not leave dependent parents, pets, or possessions that represented their lives. Others judged the danger to be insufficient to warrant evacuation compliance (Waugh Jr. 2005). Compounding these problems, absence of the Mississippi's National Guard's 155th Infantry Brigade and Louisiana's 256th Infantry Brigade hampered rescue efforts (Defense Watch 2005). Both these military units were stationed in Iraq during the Katrina disaster. On the day Hurricane Katrina hit New Orleans, a sizable portion of the Louisiana National Guard, whose troopers serve as an absolutely essential work force during any natural disaster or emergency, was 7000 miles (10,500 km) away in Iraq.

Tens of thousands took shelter in the Superdome and the New Orleans Convention Center located near downtown New Orleans. However, in the absence of necessary water, sanitation, and electricity, these places of refuge soon became uninhabitable. In the 90-degree heat, people stranded in the overcrowded Superdome were languishing. There was also no security or food. Placing people in a large shelter without a significant law enforcement presence, as well as lack of food, water, and adequate sanitation clearly contributed to the violation of human rights of those who sheltered there (Marshall 2005).

The Superdome also sustained significant damage, including two sections of the roof having been destroyed, and the dome's waterproof membrane having been peeled off (Barnshaw 2005). On August 30, Louisiana governor Kathleen Blanco ordered the evacuation of all people from the Superdome and arranged the transport for many of them to four sites: the Astrodome, the Reliant Arena, the Reliant Center, and the Convention Center all in Houston, Texas. At its peak on September 4, 2005, these four sites sheltered approximately 27,100 Katrina evacuees (Branshaw 2005). By providing shelter to as many as 150,000 Louisiana evacuees, Houston ranks at the top of US cities that received Katrina victims (Lavelle and Feagin 2006).

In their desperation, some survivors of Katrina started looting supermarkets and grocery stores throughout New Orleans to obtain food, water, ice, and other necessities. And, as to be expected, criminals among them started looting electronic shops, and departmental and jewelry stores. Reports of carjacking, murders, thefts, and rapes flooded the news. National Guard and federal troops were mobilized and sent to Louisiana along with a number of local law enforcement agents from across the country. As a consequence, some of the focus of the post-hurricane effort shifted from relief and rescue to maintaining law and order. Unfortunately, not enough law enforcement agents were available in New Orleans to maintain civil order, particularly before September 4, 2005. All, or at least most, of these events could have easily been avoided if survivors had promptly been supplied with food, water, and other essential items.

Beyond the inadequate emergency response, Hurricane Katrina has laid bare the fact of extreme poverty in the USA and also highlighted the continuing significance of race in America. For many Katrina survivors, issues of race and class were central to their evacuation experiences. It has been reported that almost 90% of all evacuees who took shelter at both evacuation sites in New Orleans and Houston were African-Americans. In obvious contrast, the law enforcement, support, and volunteer staff were predominantly white. Some evacuees reported feelings of discrimination throughout their evacuation and shelter experience (Barnshaw 2005). A survey conducted among Houston shelter evacuees by the Washington Post/Kaiser Family Foundation/Harvard University

(The Henry J. Kaiser Family Foundation 2005) reported that 68% of all respondents thought that the federal government would have responded more quickly if more people trapped in the floodwater were rich and white rather than poor and black.

A Pew Research Center poll taken approximately one week after Hurricane Katrina revealed that 66% of African–Americans believed that the government would have responded to Katrina more quickly if victims had been primarily white. In contrast, this poll revealed 70% of white respondents indicated that race of the victims had no effect on the timeliness of government response to the disaster (Pew Research Center 2005).

Six days after Hurricane Katrina tore into the Gulf Coast, the dead were still lying or floating in water in New Orleans. Corpses swelled in the sun, abandoned along the elevated highways. As rescue efforts for the living were engaged, the dead lay neglected as convoys of soldiers, police, paramedics, and boat rescue teams whizzed by.

It is clear that the US military actively participated in the post-Katrina emergency relief efforts. More than five Army and Air National Guard units comprising some 7200 active-duty troops were stationed in the Gulf Coast region to assist with hurricane relief operations. In addition, the Coast Guard and the US Navy also participated in Katrina rescue and relief efforts. As a result of the involvement of different law enforcement agents, there was confusion over, and absence of, chain of command. On September 3, 2005, National Public Radio (NPR) reported there was "no central command post" and the police in many hurricane-affected areas had no central headquarters left, no working communication systems, and few or no working vehicles. Many of the police in New Orleans quit. There had been allegations of corruption in the force before the disaster.

On a broader scale, many US States provided shelter to families and/or individuals displaced by Katrina. The majority of these survivors went to Texas, with over 230,000 taking shelter in Houston by September 5, 2005. From Texas, thousands of refugees dispersed to other states. Also, approximately 58,000 National Guard personnel were activated to deal with Katrina's aftermath, with troops coming from all 50 states. Local governments across the USA also sent aid in the form of ambulances, search teams, and disaster supplies. Schools and colleges across the country enrolled students displaced by Katrina. One thousand firefighters volunteered to assist in the relief efforts with their hometowns allocating funds to hire temporary personnel in their absence (Cooper and Block 2007). In all, a huge number of government agencies participated in the response to Hurricane Katrina, which complicated efforts to coordinate disaster relief.

Non-official Response: Domestic

Numerous nonprofit organizations, such as the American Red Cross (ARC) and the Salvation Army, provided housing, food, and water to the victims of Hurricane Katrina. The Red Cross began what it calls the largest mobilization of resources in its history for a single disaster. It dispatched hundreds of volunteers to work in disaster-impacted areas. In the first two weeks after the storm, ARC deployed 74,000 volunteers who provided shelter to 160,000 evacuees and served more than 7.5 million hot meals. It opened 700 temporary shelters for Katrina evacuees in 24 states and the District of Columbia. The American Red Cross Disaster Relief Fund collected donations from the public for their relief effort. By September 28, the ARC raised about one billion dollars in cash and pledges. Because of Red Cross efforts, monetary donations broke records set by the 2004 Indian Ocean Tsunami and 9/11 relief efforts in the USA (ARC 2010).

One important aspect of response to Katrina was that it triggered an unprecedented domestic response by nonprofit organizations that traditionally work abroad. About a dozen such organizations, including Save the Children and World Vision, worked side by side with US government agencies, churches, and charities to provide aid to hurricane victims (Wilhelm 2005). In fact, poor governmental response in some places, particularly in Louisiana, the insufficient efforts by the Red Cross prompted many American-based international relief groups to actively participate in Katrina relief efforts. Several of these groups claimed that charitable response was poorly organized, and collaboration among relief groups in Louisiana was worse than in some developing nations (Wilhelm 2005). Charities that work in both foreign and domestic disasters, however, do not agree with this assessment. They maintain that the coordination had "some bumps," but the nonprofit response worked well overall.

Another unique response to Hurricane Katrina was by Oxfam, a British relief organization, which for the first time ever responded to a humanitarian crisis in the USA.[9] The chaotic response to this disaster by government agencies and the other relief organizations involved led Oxfam to assume a new role in disaster response in the USA. ARC is the sole private organization charged by law with providing relief work immediately after a disaster in the USA (Strom 2006). This participation by international humanitarian agencies such as Oxfam occurred concurrently with the US Congress considering whether it should broaden the mandate for initial domestic disaster response beyond the Red Cross.

Many other organizations also responded in the aftermath of Katrina's devastation. Volunteers from amateur radio's emergency service wing provided emergency communications for federal, state, and local officials. Over one thousand volunteer operators traveled to affected areas to provide communication services in areas where communication infrastructure was damaged or totally destroyed. Corporations also contributed to relief efforts. On September 13, 2005, corporate donations to the relief effort were reported to be $409 million and were expected to exceed $1 billion. Wal-Mart was the largest corporate donor, contributing $26.5 million in cash and providing $3.5 million worth of products throughout Mississippi, Louisiana, and Texas (Mattera 2005).

Many criticized the media coverage of Hurricane Katrina, particularly in the city of New Orleans. Some alleged that the media spread rumors, which often magnified the problem rather than verified the facts. Some thought that the number of rapes and violence reported by the media for New Orleans in the aftermath of Katrina was inflated.[10] Conditions in the Superdome were made to seem more dangerous than necessary. As a result, the media distracted the attention of the public from what was really happening. In context of evacuation, some mainstream media and government sources chose to blame the victims as if these vulnerable people simply made bad choices, ignoring the context in which these "choices" were made (Maestas et al. 2008).

International Response

Historically, the USA has been reluctant to accept disaster aid and assistance from foreign countries other than Canada and Britain. However, as the reports of extreme damage surfaced, this policy was reversed. As many as 120 nations made pledges to provide more than $700 million in cash and other aid to Katrina victims (AP 2005). These countries included friends such as Britain and Germany and foes such as Cuba, Iran, and Venezuela. Also included were war-torn Afghanistan and countries impacted by the 2004 Indian Ocean Tsunami such as India, Thailand, and Sri Lanka. According to the US State Department (2013), Kuwait made the largest donation of $500 million, $100 million cash and $400 million in crude oil. Other large donations were made by Qatar ($100 million), India ($5 million), China ($5.1 million), and Bangladesh ($1 million).

Countries such as Canada, Mexico, Singapore, and Germany sent supplies, relief personnel, troops, ships, and water pumps to aid in the disaster recovery. Specifically, Canada sent a crew of 1600 on three naval ships, while France dispatched supplies and troops to New Orleans. Notable offers from international organizations include the United Nations, NATO, and the European Union. This latter organization sent supply water storage tanks, water purification tablets, high-energy biscuits, generators, planes, tents, blankets, meals ready to eat, and other emergency supplies along with experienced staff members.

Nearly 50 USAID experts from the Agency's Office of US Foreign Disaster Assistance (OFDA) were sent to Baton Rouge, Louisiana, to coordinate aid from foreign countries. These experts worked closely with other government agencies to make sure that the needs of disaster victims were identified clearly and accordingly transmitted to foreign governments. Vehicles from a USAID warehouse in Miami, stored there in case of natural disasters in Latin America, were used to transport non-ambulatory evacuees. USAID also provided the National Guard with latex gloves, first aid kits, hand sanitizers, and water testing kits (AP 2005).

Foreign donations sent to nonprofit organizations including the Red Cross were directed to a special fund at the State Department, or the Bush–Clinton Katrina Fund. By August 2006, the US State Department had received $126 million from 36 countries and international organizations. Many other countries, such as Canada, India, Kuwait, and Turkey, donated directly to the American Red Cross or the Bush–Clinton Katrina Fund. Unfortunately, as of August 2006, the US government had barely touched the funds donated by foreign governments (AP 2005).

Based on published information, it is clear that no one would say the federal response to Hurricane Katrina was sufficient. Even reflecting on it on the anniversary of the event, President Bush said the federal response was not sufficient. Rather, most people would use words such as "abysmal," "tragic," or "terrible." Moreover, the response was woefully slow with inadequate communication between federal and state officials and inadequate communication within the federal government itself. There was no contingency planning for evacuation of the residents of the affected areas. The response of the state government and city officials was also not adequate, but it was somewhat better than the federal response.

Kashmir Earthquake

On October 8, 2005, a 7.6-magnitude earthquake struck the North West Frontier Province (NWFP) of Pakistan, Pakistan-administered Kashmir, and parts of India and Afghanistan.[11] For its geographic coverage, it often called the South Asia Quake. The epicenter of the earthquake was Muzaffarabad, the capital of the Pakistani-controlled portion of Kashmir.[12] The epicenter was only 17 miles (25 km) away from the highly militarized Line of Control (LoC), which has divided the Pakistani-controlled part of Kashmir from India-controlled Kashmir for 58 years. The affected area covers over 11,583 square miles (30,000 square km) in impoverished, remote, and difficult-to-reach mountainous areas. The earthquake hit an area spanning at least 250 miles (375 km) from Jalalabad in Afghanistan to Srinagar in India. Essential infrastructure was destroyed, including water systems, hospitals, and government buildings. Over 12,000 schools and colleges were damaged or destroyed along with many hundreds of miles of roads and dozens of bridges (Ozerdem 2006).

Although this earthquake affected three countries, most of the damage occurred in Pakistan, particularly in Kashmir. For this reason, this disaster is also known as the Pakistan Earthquake. An estimated 87,350 people lost their lives in Pakistan, 50–60% of whom were children. This earthquake occurred in the early morning, when many children were at school. According to UNICEF, the earthquake destroyed 10,000 schools and killed an estimated 17,000 students and over 15,000 teachers attending these schools (Ozerdem 2006). In addition, more than 76,000 people were injured, and an estimated 2.8 million were left homeless (Brennan and Waldman 2006). The U.N. reported that more than four million people were directly affected by this disaster. According to official estimates, this earthquake caused approximately $5 billion in property damage in Pakistan (Ozerdem 2006).

This disaster killed at least 1360 people in India-administered Kashmir, and about 100,000 people were severely affected, and in Afghanistan, at least three persons were killed and a number of houses were damaged. Additionally, a number of high-rise buildings in the Pakistani capital of Islamabad collapsed, killing 32 people.

Global Response

The world was less generous in its response to the 2005 Kashmir Earthquake than to the 2004 Indian Ocean Tsunami. Three weeks after the earthquake, the U.N. managed to raise just 20% of the amount it was seeking compared to 80% over the same time period for the 2004 IOT. As of March 2006, only 66% of the “Flash Appeal” issued by the U.N. (an appeal for emergency aid initially estimated at $312 million and rapidly increased to $550 million) had been funded (Brennan and Waldman 2006). In response to initial calls for aid by Pakistani

President Musharraf, the world community pledged $2 billion. A month after the disaster, however, the Pakistan government had received only $9.5 million (Ozerdem 2006). Subsequently, Pakistan appealed to the world for $5.2 billion at the World Donor Conference held in November 2005. Since then, the world community and international aid agencies have pledged $5.4 billion in cash, grants, soft loans, and relief goods for rehabilitation and the reconstruction of quake-impacted areas, as well as assistance for victims (IFRC 2006).

Overseas, Muslim communities were extremely generous and quick to provide aid to the earthquake victims in Pakistan. Such communities raised large sums of disaster aid, particularly in Britain and the USA. Additionally, British volunteers of Pakistani and Kashmiri origin came to the earthquake-affected areas with both cash and needed items, such as blankets and tents (Ozerdem 2006). In Pakistan, domestic response also was overwhelming. Pakistani people from all regions and all walks of life donated huge amounts of relief goods, as well as cash, for earthquake victims. They also worked voluntarily in hospitals and in earthquake-impacted areas.

This earthquake occurred during the holiest month of Ramadan, when Muslims all over the world fast from sunrise to sunset and are especially generous to the poor and donate freely to charities. According to Ahmed Rashid (2005), even Pakistanis themselves were astounded by their own generosity. It is interesting to note that, as of the end of November 2005, an estimated $100 million was collected in donations to the President's Relief Fund. Meanwhile, private direct donations were expected to be five to ten times that amount (Rashid 2005).

Initially, however, inadequate response from the international community showed itself in two ways: in not meeting initial aid pledges made immediately after the disaster and in delays in providing a sufficient number of helicopters necessary to support the aid effort in the remote mountainous regions. In addition to the rugged mountainous topography in the Kashmir region, roads were blocked by landslides associated with the earthquake and its aftershocks. In this situation, helicopters were the only effective means to reach earthquake victims. Unfortunately, offers of additional helicopters to Pakistan were limited and delayed. As a result, Pakistan used its 15 helicopters heavily in the early days of relief operation (Ozerdem 2006).

The paucity of funds and the pace at which they were released also slowed down the relief operation affecting how quickly people were reached, how quickly relief supplies were delivered, and more importantly, the efficiency of the operation. There were a number of reasons for an initial weak and tardy global response to the Kashmir Earthquake. Unlike the 2004 Indian Ocean Tsunami, Western tourists were not directly affected by this disaster. Moreover, it occurred in early October and not Western holiday period like Thanksgiving or Christmas. By the time this disaster occurred, individuals and governments who would ordinarily contribute to such a serious cause had become jaded from the constant demand for support produced by recent disasters. Relief agencies and volunteers call this, "donor fatigue," "donation fatigue," or "compassion fatigue." Many felt as though they had given enough, while others were uncertain about how their funds would be utilized.

Some relief experts maintain that a multi-country event, such as the 2004 Indian Ocean Tsunami, may attract more support from donor countries than a single-country event. This might help explain why the world has not responded to the earthquake as generously as to the 2004 tsunami. Additionally, some donor countries, such as the USA, suffered their own disasters in 2005. Additionally, lack of confidence in the U.N.'s ability to respond was often cited as a reason for the poor response received to its appeal for emergency aid to Pakistan. Moreover, prior to this disaster, print and electronic media broadcast reports of the U.N. scandal regarding the Food For Oil program in Iraq, which damaged the reputation of the U.N. (see Otterman 2005).

The delayed and inadequate initial response from the USA was a shock to many Pakistanis and government officials since the USA is on the front lines in the war on terrorism. However, apart from Iraq and Afghanistan, no country is more critical of the war on terror than Pakistan. At any rate, the USA pledged only $50 million in aid. Turkey, in contrast, offered $150 million in aid, becoming the largest donor for Pakistani quake victims (Ozerdem 2006). Surprisingly, India was the first country to offer help. In turn, Pakistan accepted Indian assistance only in the form of food, medicines, and tents. In all, 67 nations and numerous foreign relief organizations responded to the Kashmir Earthquake. Relief operation included distributing more than 130,000 tons of food and giving shelter to 150,000 people in 474 tent cities.

Participation of Foreign Military

Military forces from several Western countries included in the North Atlantic Treaty Organization (NATO) took part in disaster relief efforts and provided necessary helicopters for carrying the wounded to hospitals, transporting search and rescue teams, and delivering relief assistance. However, participation of foreign forces in disaster relief efforts was criticized by both opposition parties in the country and Islamist groups. At the same time, despite the desperate situation, the Pakistani government refused an offer of Indian army helicopters because India would not allow Pakistani pilots to fly them. This is what opposition parties and Islamist groups were concerned about the presence of foreign forces in the country.

Earthquake Relief and LoC

Ultimately, the earthquake in Kashmir provided a unique opportunity for both Pakistan and India to build trust and take some political steps toward a peaceful resolution of the Kashmir conflict. Both sides took several steps toward the smooth flow of disaster aid to Kashmiri people living on both sides of the LoC. India lifted restrictions over mobile phone connections between families separated by the 500-mile-long (742-km-long) LoC. Cross-border phone links were disconnected and have remained so for the last 16 years. Within two weeks of the disaster, India had delivered close to 300 tons of food, tents, and other essential supplies to Pakistan. India also consented to allow Pakistani helicopters to operate in a no-fly zone along the LoC (Renner and Chafe 2007).

The governments of both countries opened five crossing points along the LoC starting on November 7, 2005, which facilitated exchange of humanitarian assistance and allowed separated families to meet. This represented a possible starting point for a series of confidence-building measures. These attempts may produce some positive outcomes, as has been the case in a number of other disaster-affected countries. For example, after the Marmara Earthquake, the relationship between Turkey and Greece improved. In Aceh in Indonesia, the 2004 tsunami created a fertile environment for a peace agreement between the Indonesian government and the Free Aceh Movement (GAM). The same is true for Sri Lanka, where Tamil separatist forces have been fighting for independence for more than two decades (Renner and Chafe 2007).

Participation of National Army

Despite warnings in recent years of potential seismic activity, Pakistan was ill-prepared for a disaster of this magnitude having no formal mechanism for disaster preparedness and no governmental department dedicated to handling disaster relief. In fact, the last earthquake of this magnitude occurred as long ago as 1935 in Quetta, killing 30,000 people (Ozerdem 2006). Thus in 2005, the army was the obvious option to lead and organize the huge response required to meet the needs of the earthquake's victims (IRIN Pakistan 2006; IRIN 2010). Accordingly, following the earthquake, General Pervez Musharraf decided that emergency relief efforts would be a military-run affair and dispatched 50,000 troops into affected areas.

Demands from opposition parties, and from some donors, for more civilian and local oversight were brusquely turned down because neither civilian authorities of the Pakistani government nor local authorities were prepared to deal with the devastation caused by this earthquake. For instance, local authorities were paralyzed by the deaths of many of their employees and/or experienced deaths in the family. Therefore, the army was the only institutional framework that the government could come up with to respond to this disaster. Even the army, however, was not prepared for such a task because it lacked appropriate training and proper equipment to find and extricate the many survivors from the rubble.

Pakistan's institutional response to the Kashmir Earthquake is referred to by Ozerdem (2006) as a military-centered and conveyed approach, in which President Musharraf was the Chief-of-Staff of the process. Further, many claim that military response to the Kashmir Earthquake was slow and ineffective because of the state's lack of preparedness as well as the military waiting for top-down orders that never came. Compounding the situation, landlines, radio communication, and mobile phone systems were severely damaged by the earthquake (McGirk 2005). The military was also unwilling to coordinate its activities with civilian authorities.

To some large degree, though, Pakistani troops restored order, cleared landslide-affected roads, replaced bridges, and re-established communications in the difficult mountainous environments. They helped local people search for those trapped, and conducted household surveys to establish the extent of damage and to assess the levels of need for disaster victims. In sum, the troops evacuated over 80,000 people from the affected areas (IRIN Pakistan 2006). The Pakistani army also worked closely with foreign military personnel who came from dozens of countries during the relief phase.

At the height of the emergency, more than 100 international and national aid agencies, including foreign and domestic NGOs, were active in the quake areas (*The Economist* 2006). Rescue and relief workers from different parts of the world brought rescue equipment, including helicopters and rescue dogs. Before the catastrophic earthquake, foreigners were not allowed to travel to or through Kashmir. This disaster has opened a window of opportunity that was otherwise rare.

The most spectacular aspect of the relief operation was the presence of the numerous local and Islamist groups, working side by side (though not necessarily cooperating with each other). Participation, particularly from the local Islamist groups and their extremist cousins, to aid relief efforts seems to have much wider ramifications for the sociopolitical landscape of Pakistan. Some maintain that the army's slow and largely ineffective response to the disaster created a great opportunity for local religious groups to win people's hearts and minds through the deployment of large and effective resources to the disaster (Rashid 2005).

Local Islamist groups, notably the Jamaat-e-Islami (JeI) and the Jamaat-ud-Dawa (JuD), established field camps, tent villages, food centers, and mobile field hospitals for the earthquake victims throughout the affected areas.[13] Much of the quake-affected area is home to a conservative Muslim society. By using their local networks and plentiful funds, disciplined cadres of these religious groups helped people when it was needed most, indirectly eroding the credibility of the army and President Musharraf. The quake-affected people of Pakistan-administered Kashmir were touched and impressed by the efforts of the Islamists. Various Islamic groups tried hard to be the first with aid and posed a serious threat to President Musharraf's regime (Qureshi 2006).

Ultimately, together, the army and others involved in the disaster relief efforts in Kashmir established 144 camps, built 425,000 emergency shelters, distributed 520,000 tents, and immunized 350,000 children against measles. Although most of the tents were not suitable for harsh winter conditions, and many collapsed under heavy snow, still the people survived the winter with fewer deaths from cold than were recorded the previous year. Before the arrival of winter, there had been a severe shortage of winterized tents, sleeping bags, and blankets. It was projected that a "second wave" of deaths was a real possibility in the earthquake-impacted areas unless global response immediately occurred, which it did.

The emergency aid Pakistan ultimately received from the rest of the world was not insignificant, but it was well short of the global response to the 2004 Indian Ocean Tsunami, in both its speed and its amount. Also, it appears that the aid Pakistan received has been well spent. The Pakistani army seems to have surprised everyone, including most Pakistanis, and there's no question that, without their hard labor, the death toll would have been higher (Renton 2005). U.N. Secretary General Kofi Annan's warning of a "second massive wave of death" did not materialize because of the international community's response and the works of many volunteers as well as members of the Pakistani army.

Not much information is available regarding the emergency response to the 2005 Kashmir Earthquake that occurred in the Indian-administered Jammu and Kashmir regions. However, a few details have emerged. For instance, the Indian government did appeal for external aid for earthquake victims and reports indicated that the injured were treated in hospitals and in makeshift medical facilities. In keeping with its traditional duty since independence, the Indian army also undertook a key role in coordinating and running relief operations.

2010 Haiti Earthquake Response

On January 12, 2010, at 4:53 p.m., a 7.0-magnitude earthquake struck Haiti. This was the largest earthquake ever recorded in Haiti (Kirsch et al. 2012). The epicenter of the earthquake was 13 miles (20 km) west of the country's capital, Port-au-Prince. The major event was followed by up to 59 aftershocks, each registering greater than 4.5 on the Richter scale (Coles et al. 2012). Estimates suggest that approximately one-third of its nine million people have been affected by the earthquake. More than 300,000 homes collapsed, leaving unknown numbers of people trapped; further, more than 700,000 people were displaced, and many people were either evacuated from Port-au-Prince or left the capital on their own for neighboring cities and rural areas not affected by the earthquake. This event caused estimated deaths tolls of 222,750 and injured 300,000 others (Kirsch et al. 2012). One million children either lost one or both parents (Hebblethwaite 2011). Overall losses and damages from the earthquake were estimated to be between $7 billion and $14 billion (approximately 100–200% of Haiti's gross domestic product) (Cavallo et al. 2010).

In the immediate aftermath, the President of Haiti made an appeal for $575 million assistance, and within six months, the country received more than 100% of its appeal amount (Kirsch et al. 2012). The humanitarian response to the 2010 earthquake in Haiti was one of the largest in history, and this appeal unleashed an unprecedented flood of emergency aid: $13.5 billion in donations and pledges (Knox 2015). However, the massive humanitarian relief operation in the country has experienced a huge range of significant challenges that reduced the effectiveness of this operation. These challenges, among other things, were created because of damaged infrastructure, which, in turn, delayed distribution of emergency supplies. For this reason, some affected areas had not received any emergency relief 10 days after the earthquake (Hebblethwaite 2011). Other problems included an outbreak of cholera, and a lack of appropriately experienced health professionals and poor coordination between government, military, and non-governmental organizations involved in the relief operation (Margesson and Taft-Morales 2010).

Following the earthquake, the US Ambassador to Haiti issued a disaster declaration, and the US Agency for International Development (USAID), through the Office of Foreign Disaster Assistance (OFDA), authorized $50,000 for the initial implementation of an emergency response program. The next day, the USA announced it would send $100 million in humanitarian assistance to Haiti to meet the immediate needs. Also, the US Department of Homeland Security temporarily halted the deportation of Haitians and granted Temporary Protected Status for 18 months to Haitian nationals who were in the USA as of January 12, 2010 (Margesson and Taft-Morales 2010). Notably, an estimated 45,000 US citizens were living in Haiti at the time of the earthquake (USDHS 2010). With the help of the US embassy, more than half of them returned to the USA (USDHHS 2010).

Prior to the earthquake, the international community, including different agencies of the United Nations and World Bank, had been working in Haiti for decades to provide extensive development and humanitarian assistance to the country. These organizations were also working with the Haitian government to implement disaster risk management into the country's overall development strategy and to develop its capacity for disaster response. The capacity was still in its early stages, however, and the focus of much of its risk management efforts was not on earthquakes, but on hurricanes, which are the most common natural disasters on the island. In 2008, more than 800 Haitians died in a series of four hurricanes and tropical storms that struck the country during a two-month period (DesRoches et al. 2011). However, with that assistance from various external sources, the Haitian government had made significant progress prior to the earthquake in recovering from those devastating storms.

The United Nations, which already had a strong presence in Haiti, was at the forefront of response for security and humanitarian assistance. The organization itself suffered heavy losses as well. Its headquarters collapsed, and about 150 U.N. personnel were unaccounted for (Margesson and Taft-Morales 2010; Hebblethwaite 2011). Notably, the U.N. Security Council created the U.N. Stabilization Mission in Haiti (MINUSTAH) on April 30, 2004, and since then, International Peace Keeping Forces have been stationed in the country with a mandate to focus on three broad areas: a secure and stable environment, and improving the political process, and monitoring human rights (Margesson and Taft-Morales 2010). In addition to U.N. agencies, prior to the earthquake, about 10,000 NGOs workers were in Haiti primarily to alleviate poverty (Versluis 2014). After the earthquake, this number increased significantly, and they received the bulk of the global relief funds to distribute to earthquake survivors. In 2010, Haiti had the second highest number of NGOs per capita in the world. For this reason, Haiti is often considered "a Republic of NGOs" (US Institute of Peace 2010) or "The NGO Republic of Haiti" (Klarreich and Polman 2012).[14]

Haitian Government and U.N. Efforts

Immediately after the earthquake, the highest priority of the Haitian government was to conduct search and rescue operations for survivors. Other priorities included an offshore vessel medical unit and electricity generation capability. Within a day of the earthquake, the USA deployed search and rescue teams along with support staff and equipment. Within the first 12 days after the earthquake, members of 43 domestic and foreign SAR teams had rescued 134 people (Margesson and Taft-Morales 2010). They also helped "the U.N. Children's Fund (UNICEF) to organize tent and equipment donations to establish child-friendly spaces and health facilities" (Margesson and Taft-Morales 2010, 2).

After the earthquake, survivors, including those who had lost their homes, were frightened to sleep indoors and began to gather spontaneously in available open spaces (e.g., parks, roads, soccer fields, a golf course, and other open spaces) in Port-au-Prince (Versluis 2014). They set up temporary shelters/camps with plastic sheets, cardboard, and tarpaulins. To provide better roof material for these displaced people, the Haitian government made a donation appeal for 200,000 family-sized tents (Margesson and Taft-Morales 2010). Emergency shelters were also needed for nearly half a million survivors who were displaced outside Port-au-Prince. Subsequently, aid workers delivered basic necessities to all emergency shelter locations. The World Food Program (WFP) initially provided two-week rations to two million people in Port-au-Prince through a new fixed distribution site system, and OCHA reports that the Government of Haiti provided food kits to 100,000–150,000 people per day (Margesson and Taft-Morales 2010).

To coordinate the assistance delivered by numerous humanitarian organization in Haiti, the U.N. formed 12 clusters, which are groups of humanitarian organizations, both U.N. and non-U.N., in each of the main sectors of humanitarian action (Table 4.4). These clusters were designated by the Inter-Agency Standing Committee (IASC) and had clear responsibilities for coordination (Margesson and Taft-Morales 2010). The Cluster Approach was applied for the first time following the 2005 earthquake in Pakistan (Street and Parihar 2007). Each cluster was led by the agency listed in the third column of Table 4.4. These agencies were responsible for coordinating the provision of aid, and each cluster coordinated with the U.N. Office for the Coordination of Humanitarian Affairs (OCHA) (HHI 2011).

On January 15, 2010, a U.N. team stationed in Haiti at the time of the earthquake made a Flash Appeal of $575 million for emergency aid to the survivors of the event. By February 1, 83% of the Flash Appeal had been received and the remaining 17% had been pledged (Margesson and Taft-Morales 2010). Along with many other countries, the USA provided assistance in the form of direct contributions of relief goods or through the operation of relief flights and logistics support. In addition to assistance provided by bilateral and multilateral donors, funding was provided by governments of other countries (Margesson and Taft-Morales 2010), particularly Caribbean Community (CARICOM) countries.

Private companies and individuals also contributed more than $470 million to support relief efforts in Haiti. Additionally, private initiatives

Table 4.4 Twelve clusters for the 2010 Haiti relief operations

No.	Sector name	Lead agency
1	Agriculture	Food and Agriculture Organization (FAO)
2	Camp Coordination and Camp Management	International Organization for Migration (IOM)
3	Early Recovery	U.N. Development Program (UNDP)
4	Education	U.N. Children's Fund (UNICEF)
5	Emergency Shelter and Non-Food Items	IOM and the International Federation of Red Cross and Red Crescent Society (IFRC)
6	Emergency Telecommunications	World Food Program (WFP)
7	Food	WFP
8	Health	World Health Organization (WHO) and Pan American Health Organization (PAHO)
9	Logistics	WFP
10	Nutrition	UNICEF
11	Protection	Office of the High Commissioner for Human Rights (OHCHR)
	Child Protection	UNICEF
	Gender-Based Violence	U.N. Population Fund (UNFPA)
12	Water, Sanitation and Hygiene	UNICEF

Source Prepared by the author based on information provided by Margesson and Taft-Morales (2010), p. 8

in the USA, such as the campaign by the American Red Cross to raise funds through text messages ($29 million), and numerous local fund-raising activities increased private giving. On January 16, 2010, former Presidents Bill Clinton, who was also serving as the U.N. Special Envoy to Haiti, and George W. Bush led a fund-raising effort and worked with the US private sector in support of Haiti. The initiative is called the Clinton Bush Haiti Fund (http://clintonbushhaitifund.org). This nonprofit organization was founded in response to the request of President Obama to assist in the Haitian rescue effort, and it encouraged cash donations.

After the earthquake, the World Bank announced $100 million in emergency grant funding to support recovery and reconstruction, but this international financial institute did not provide grant for emergency relief. On January 12, 2010, the day of the earthquake, the

Inter-American Development Bank (IDB) announced a $200,000 emergency grant for immediate relief aid (Margesson and Taft-Morales 2010). Haiti also received concessional loans from the International Monetary Fund (IMF) as well as from the multilateral development banks.

US Humanitarian Assistance

In response to the crisis in Haiti, the US Department of Defense (DOD) deployed a broad range of military assets. Thus, in early February 2010, there were 20,458 US military personnel in Haiti. Additionally, 26 US Navy and Coast Guard ships were on site, and 68 helicopters and over 50 fixed-wing aircraft assisted in the transportation of relief goods and casualties. Within a day of the earthquake, the US Air Force Special Operations Command personnel were deployed to the Port-au-Prince International Airport to restore air traffic control capability and to enable round-the-clock airfield operations (Margesson and Taft-Morales 2010). In fact, humanitarian supplies from abroad came to Haiti via either Port-au-Prince or Santo Domingo, Dominican Republic. The airport in the Dominican Republic was used as a humanitarian staging area to help with the coordination effort and allow for relief teams and supplies to get to Haiti by land. There had been some concerns about security and potential for looting and violence, but the overall situation remained calm and stable, with only sporadic incidences of looting and criminality (Margesson and Taft-Morales 2010).

Margesson and Taft-Morales (2010) reported that as of February 1, 2010, US military forces delivered 2.1 million water bottles, 1.79 million food rations, more than 100,000 lbs of medical supplies, and more than 844,000 lbs of bulk fuel. In addition, personnel of the DOD provided many services to the earthquake survivors, including "casualty treatment both ashore and afloat, aerial reconnaissance to assist rescue/supply efforts, the distribution of hand-held commercial radios, and the provision of radio broadcast capacity for emergency services information" (Margesson and Taft-Morales 2010, 12). Furthermore, the

US Navy deployed 19 ships to assist relief efforts, while Navy surveillance aircraft performed initial aerial surveys of the earthquake damage.

Along with numerous other NGOs, USA-based NGOs played an active role in the relief effort in Haiti. Several of the USA-based NGOs were funded by the US government. Similar to that of the government, the response from US households was overwhelming: a staggering 50% of all US households donated to the Haiti recovery cause (Pew Research Center 2010).

Disaster Relief Aid

There is no reliable estimate about what proportion of earthquake survivors in Haiti stayed in camps. Anna Versluis (2014), who conducted interviews approximately seven weeks after the earthquake among five camps, reported that 32% of the camp dwellers had not received any material assistance from any sources since the disaster. She further claims that excluding water, emergency assistance from relief agencies and organizations had reached only 55% of the households participating in her study, and the amount of assistance they received was modest. Versluis (2014) suspected that those not living in camps probably received even less aid. This finding is consistent with Kirsch et al.'s (2012) observation that non-camp residents were more likely to receive less humanitarian assistance than camp residents. One possible explanation for this difference was that members of the former group were economically well-off relative to the members of the latter group.

Another study (Griffin et al. 2010) conducted in late February 2010 reported that 64% of earthquake survivors received emergency assistance from aid agencies. Several months after the earthquake, Schuller (2010) conducted a study and reported that 53% of all camp residents received disaster aid. This study estimated that 41% of the displaced persons received a tent or tarpaulin from relief agencies, and this finding corresponds exactly with that of another study (Bigg 2010). These findings suggest that many basic needs of people in the camps were not being met. Also, Versluis (2014) claimed that emergency aid received from social networks played a principal role in the 2010 Haiti

Earthquake relief efforts. In contrast to in-kind aid from relief agencies, survivors received cash donations from friends and relatives, living in both Haiti and abroad. The cash transfers were timely more often than aid-in-kind provided by the relief agencies (Versluis 2014).

In addition to slow response, the relief goods received by the earthquake survivors in Haiti were inadequate, and many aid materials were not useful for emergency situations. For example, "about 80% of clothing donations were useless" (Holguin-Veras et al. 2014, 4). Moreover, humanitarian aid agencies experienced numerous problems in disbursing relief goods to those in need: shortage of labor and transport vehicles, lack of security and local knowledge needed for such a relief effort, very few points of distribution, and poor coordination. Many aid agencies bypassed local and national authorities to deliver earthquake relief, and they made decision about relief distribution without consulting key Haitian government authorities (Oxfam 2011). While preexisting social networks did very well in providing disaster aid, newcomers struggled and could not properly deliver the supplies (Holguin-Veras et al. 2012).

Widespread use of social media (e.g., Twitters, Blogs, Facebook, and Web sites) including crisis mapping by OpenStreetMap (OSM) in post-disaster situations started because of the 2010 Haiti Earthquake and the media contributed more to the rescue and relief operations in this country than they did to any previous operations (Slagh 2010; Seto and Nishimura 2016). In fact, social media also played a critical role in raising funds for the disaster survivors, expanding volunteerism opportunities, and increasing Haiti's access to news coverage. Celebrities used social media to ask their fans to donate for the disaster survivors and affected communities in Haiti. In the post-disaster relief phase, a large number of volunteer and technical communities (V&TCs) for the first time mobilized tools for collecting, analyzing, visualizing, and distributing data and maps among responders, and field staff of NGOs and U.N. agencies (HHI 2011). Since the Haiti Earthquake event, social media have become effective tools in enhancing the ways in which traditional disaster relief operation is implemented (Slagh 2010).

Despite such progress, humanitarian response in Haiti suffered from a range of different problems that reduced the effectiveness of relief operations. Notably, there was a lack of cooperation among the agencies and individuals involved in managing and distributing disaster relief as well as a lack of relevant experience particularly among the members of over 600 health agencies who responded to the Haiti Earthquake. For example, one internet source (GWWC 2014) reported that very few of the health agencies had the competence to provide appropriate surgical care. A lack of coordination resulted in providing the same emergency care to the same area, while other affected areas were left with no access to emergency care. For coordination problems, after a few weeks, many health agencies left Haiti, leaving behind thousands of post-surgical patients.

Another reason for emergency medical responders leaving the country was large delays in getting in place the medical equipment that was urgently needed to treat injured victims of the earthquake. Largely, the delay was caused by airport congestion as the airport in Port-au-Prince gave dignitaries and media priority, while planes carrying humanitarian supplies were diverted to Santo Domingo (the capital of the neighboring Dominican Republic), and then transported the 200 miles to Port-au-Prince by land (GWWC 2014). This delay significantly delayed relief efforts.

Often, confusion was created among the participating groups and individuals as to over who was running the relief effort: the USA, which controlled the Port-au-Prince airport, the U.N., which was overseeing distribution, or Haitian President Rene Preval who claimed he was in charge of relief distribution. A report from the Guardian maintained that Preval's role was largely limited to appealing for assistance and meeting visitors (McGreal and Addley 2010). Sufferers of this confusion were the earthquake survivors, a sizable number of them still living in camps.

Famine in South Sudan

On July 9, 2011, South Sudan, officially the Republic of South Sudan, was born peacefully through a referendum as the 54th independent country in Africa.[15] However, this peace did not last long. On December 15, 2013, an armed confrontation erupted in the capital of

Juba between army officers loyal to President Salva Kiir and disgruntled soldiers backing his ex-deputy Riek Machar that soon began to deteriorate into a civil war (Maru 2013). Earlier, in July, President Kiir dismissed the entire cabinet, including Vice President Machar. The confrontation has been brewing since its birth along the ethnic lines of the president and former vice president who are from two distinct tribes (Maru 2013). Their respective tribes' mistrust of each other immediately caused each leader to accuse the other of political oppression and to promote a dictatorship. Moreover, the government was corrupt and had managed poorly its resources and economy.

President Kiir invited Ugandan military forces to fight alongside South Sudanese government forces against the soldiers loyal to Machar (Museven 2014). Meanwhile, the United Nations deployed peacekeepers in the country as part of the United Nations Mission in South Sudan (UNMISS) to protect civilians, to support the government in developing its capacity to provide security, and to establish rule of law. Numerous cease-fires were mediated by the Intergovernmental Authority on Development (IGAD) between two factions of the Sudan People's Liberation Movement (SPLM) and subsequently failed. Thus, a peace agreement was signed in Ethiopia under threat of United Nations sanctions for both sides in August 2015. Nevertheless, Machar returned to Juba from exile in April 2016 and was once again appointed vice president. Following a second breakout of violence in Juba, Machar was ousted in July, and he fled the country as the conflict erupted again.

Three years of civil war has devastated the country. Unity state in the northern-central part of South Sudan, which borders Sudan, has been at the center of some of the fierce fighting, and overall, this civil war has killed up to 300,000 people (News24 2016). Although both leaders have supporters from across South Sudan's ethnic divides, subsequent fighting has been communal, with rebels targeting members of Kiir's Dinka ethnic group and government soldiers attacking Machar's Nuers group. More than 3.5 million people have been forced to flee their homes in a country of 12 million, with about 2 million internally displaced and about 1.5 million having fled to neighboring countries, especially Kenya, Sudan, and Uganda (AP 2017).

Famine Declaration and Response

The civil war has severely disrupted food production in South Sudan where up to 95% of the country's population depends on farming, fishing, or herding to meet their food needs. The early part of 2017 brought a severe drought, which has continued in the southeast part of South Sudan (Cook and Robins-Early 2017). Thus, conflict and drought have been forcing people to abandon their homes and their lands. As agricultural seasons have been repeatedly missed and livelihoods abandoned, the number of people on the brink of famine in South Sudan has been rising. Correspondingly, crop production has plummeted since 2013. Given this grave situation, a famine was declared in February 2017 in parts of South Sudan in what the U.N. described as a man-made catastrophe caused by civil war and economic collapse (Quinn 2017a). According to the World Food Program (WFP), about 100,000 people in the country faced starvation, and an additional one million people were classified as being on the brink of famine in the early part of 2017. At that time, about 4.9 million people, more than 40% of South Sudan's population, were desperately in need of food (Quinn 2017a).

The U.N. famine declaration has not received much attention from bi- and multilateral humanitarian organizations beyond the stated agencies, NGOs, faith-based organizations, and others working within the country for the last several years. In response to this famine declaration, the WFP has intensified its emergency food distribution in famine-affected parts of South Sudan. The organization has already been providing food to people sheltering in U.N. bases and in refugee camps. Other U.N. agencies (e.g., FAO, WHO, and UNICEF) have renewed their efforts to provide food, seeds, livestock, and health care to people in famine-prone areas of South Sudan.

As noted, several organizations, such as Episcopal Relief and Development, have been active since 2014 in South Sudan to support local efforts to provide food aid through its partners and by working with relief agencies and diocesan leaders in the country. The organization has a continuing presence in areas that, while not meeting the definition of famine until now, have long been dealing with extreme food shortages (Paulsen 2017). After the U.N.'s famine declaration, the organization expanded its support of relief efforts in South Sudan. Similarly, Mercy Corps has been on the ground since before the famine declaration. After the declaration, this organization reinforced its efforts to help children and families by providing food and other necessary food items.

Doctors Without Borders has been working in what today constitutes the Republic of South Sudan since 1983 despite five Sudanese staff members of this organization being killed in Unity state. Members of this group have been providing basic health care through medical centers and mobile clinics. This organization also responded to outbreaks of diseases in refugee camps. After the U.N.'s famine declaration, this organization extended its program to include distribution of food in famine-affected areas of South Sudan.

In response to the U.N. declaration of famine in South Sudan, the International Federation of the Red Cross and Red Crescent Society started to raise funds to supply food for the starving people of South Sudan. Among all countries of the world, the U.K. seems to be the first major donor to respond to the U.N.'s appeal (Quinn 2017b). The British government announced a donation of £100 million that had, in reality, already been reserved for the war-torn country. Most of the media outlets in Britain wrongly interpreted this donation as providing life-saving food, water, and emergency health care. The U.K.'s response was followed by that of the EU who pledged £69 million to tackle famine in South Sudan (Quinn 2017c).

Although the USA has been the largest donor of humanitarian assistance to South Sudan since 2014, the country accused South Sudan of carrying out "deliberate starvation tactics" and claims that the famine is not a result of drought but the result of political control and personal gains by the political leaders of both groups (Iaccino 2017). After the famine declaration, the Government of South Sudan obstructed access to aid workers and increased work-permit fees for foreign aid workers from $100 to $10,000 (Iaccino 2017). Additionally, Kiir's government has been accused of looting aid supplies (Cook and Robins-Early 2017).

A U.N. declaration of famine was supposed to bring a surge of assistance to South Sudan. But it did not happen for many reasons. Within days of the declaration, the South Sudanese government ordered aid workers and their local partners to leave ahead of a planned offensive, and the area was soon consumed by fighting. Thus, the workers were blocked from reaching the needy, in some cases by insurgents, in others by soldiers loyal to the president (Sieff 2017). Apart from such obstruction and security concerns, workers faced serious logistical challenges reaching families in need.

The country's formidable weather and lack of infrastructure mean transporting humanitarian aid can be incredibly difficult. During the rainy season, when parts of South Sudan become flooded and the roads are impassable, the WFP and other humanitarian organizations rely on strategically pre-positioned food stocks and airdrops to reach the most remote communities. It is suspected that South Sudan's lean season, the months before harvest time when food stocks and job opportunities dwindle, will hit in July and push even more families into poverty and hunger (Altman 2017).

In the absence of meaningful peace and security, new humanitarian and relief organizations are not likely to enter South Sudan. Even those within the country cannot provide necessary assistance to people of the crisis-affected areas. Besides, the forces loyal to both factions have been attacking the premises and warehouses of U.N. agencies and other relief organizations (AP 2017). Likely, the famine in South Sudan will be reduced only when there is political stability in the country.

Conclusion

Among the five major disasters considered in this chapter, response, in general, and international response, in particular, was unprecedented for the 2004 Indian Ocean Tsunami. That response was quick and timely. The massive tsunami during a December holiday, complete with widespread, real-time media coverage, was doubtless the catalyst for the subsequent huge outpouring of sustained aid.

Rapid response has not been the case in the 2017 South Sudan famine and was not the case after the 2005 Kashmir Earthquake. For the Kashmir Earthquake, it took a considerable amount of time for the international community to respond. However, domestic response, particularly from individuals and voluntary organizations, earned praise from Kashmir Earthquake survivors despite delays because of remoteness, rugged terrain, and weather conditions of the affected areas.

Regarding the Pakistan Earthquake, there was dissatisfaction among opposition political parties for involving its military in emergency relief efforts. It appears, however, that public agencies in Pakistan were not prepared to perform this massive task.

Finally, emergency relief efforts after Katrina were considered by many as anything but prompt and unbefitting the magnitude of the disaster. Newsweek reported that a "strange paralysis" set in among Bush administration officials, who debated lines of authority while thousands died (Thomas 2005). This centralized decision making caused delays in critical decisions, such as the deployment of National Guard units and mismatches between local needs and national actions. A more decentralized approach would have saved lives, time, and suffering for many victims. Similarly, better coordination in Haiti would have reduced the tremendous sufferings of the 2010 earthquake survivors.

However, Hurricane Katrina received some of the highest amounts of emergency aid of the five disasters considered in this book. The volume of media coverage in Katrina's aftermath, including criticisms of the inadequacy of the government's response, spurred more donations than did previous disasters. Americans' familiarity with New Orleans as a tourist destination and known cultural symbols drove a lot of the support. This implies that race and low income did not alone slow fund-raising. Unlike for the four other disasters, the aftermath of Katrina initiated a "blame game" to cover up the inability of the US federal government to respond quickly and appropriately, particularly in a city the size of New Orleans. Federal agencies blamed state and local agencies including the Louisiana government and the New Orleans mayor while former FEMA director Michael Brown blamed the victims in flooded New Orleans for their dire situation because they did not leave.

The scale of four of the five disasters considered in this book was so large that all affected countries deployed their national and/or foreign military force, and almost all of them including the USA allowed foreign forces to participate in rescue and disaster relief operations. They also provided the necessary infrastructure that facilitated the delivery of emergency aid in a timely manner. However, in the USA, the appropriate role for the military has become a major issue. People debated whether the military should become the nation's disaster relief organization, in addition to its mission to defend the nation from external threats. This also raised questions as to whether or not the National Guard should be used to support regular forces overseas given the inability of National Guard personnel to be deployed to Katrina-affected areas because many units were overseas. Lessons learned from the five disasters considered in this book can be used to improve relief operations for future extreme events both natural and man-made.

Notes

1. The magnitude of this earthquake was originally recorded as 9.0 but has been upgraded to between 9.1 and 9.3 (Paul and Nadiruzzaman 2013).
2. These countries are Bangladesh, India, Indonesia, Kenya, Malaysia, Myanmar, the Maldives, Myanmar, the Seychelles, Somalia, Sri Lanka, South Africa, Tanzania, and Thailand.
3. Many sources reported the number of deaths, but the death figures differ from one source to another. The number generally increases as the length of reporting time increases from the date of occurrence of the event.
4. Apart from the U.N., several other organizations also launched appeals. For example, in the UK, the Disaster Emergency Committee (DEC) also made an appeal for the IOT.
5. Table 4.2 is prepared from Table B of the Financial Tracking Service of UNOCHA (http://fts.unocha.org; last accessed on 31 December 2016). It seems that Table B excluded several funding sources.
6. For other reasons regarding Japan's commitment to provide emergency aid to tsunami victims, see Tamomoto (2005).

7. It is important to note that response to natural disasters is easier and less politically charged than complex emergencies like the one in Darfur, Sudan. Moreover, news and other electronic media do not have the same access to areas under conflict, nor are events in these conflicts usually as sudden or as massive as natural disasters (Spiegel 2005). Additionally, researchers note relief effort loses its momentum with reduced media attention.
8. Problems associated with 2004 tsunami relief operations were also discussed in Chapter 3.
9. More than one dozen INGOs (e.g., International Rescue Committee, International Medical Corps, and International Relief and Development) provided significant humanitarian assistance and relief for the first time ever in the USA (Eikenberry et al. 2007).
10. Some claim that reports of rape were untrue. For example, Matt Welch writes on Alernet that, "the rape reports were 'invented' out of whole cloth" (cited in Marshall 2005, 15).
11. The NWFP is now called Khyber Pakhtunkhwa (KPK) province.
12. The heaviest damage occurred in Muzaffarabad where an estimated 32,335 buildings collapsed.
13. The JuD is the political arm of the banned extremist militia Lashkar-e-Taiba (LeT), familiar for its pro-Taliban and pro-Kashmir jihad views. Some of its militant training camps in Pakistan-administered Kashmir are said to have been destroyed.
14. Other sources (e.g., Edmonds, n.d.) claim that Haiti ranks top in the world with respect to NGO per capita.
15. It became an independent country after over 20 years of guerrilla warfare against the government forces of Sudan. The warfare claimed the lives of at least 1.5 million people, and more than four millions were displaced. South Sudan has an area of 239,285 sq. miles (619,745 sq. km). Made up of the 10 southern-most states of Sudan, South Sudan is one of the most diverse countries in Africa. It is home to over 60 ethnic groups (Nijman et al. 2016).

References

Altman, M.J. 2017. 5 Things to Know About Famine in South Sudan. WFP. https://wfpusa.org/articles/5-things-to-know-about-famine-in-south-sudan/. Last Accessed 1 June.

AP (Associated Press). 2005. Foreign Aid to U.S. for Katrina Relief. 10 Sept. http://www.nbcnews.com/id/9282598/ns/us_news-katrina_the_long_road_back/t/foreign-aid-us-katrina-relief/#.WTV4zesrJhE. Last Accessed 2 Jan 2008.

AP (Associated Press). 2017. UN Food Agency Warns South Sudan Conflict Is Fueling Famine. May 30. https://www.yahoo.com/news/un-farm-agency-south-sudan-avoid-famine-help-124239445.html. Accessed on 31 May 2017.

ARC (American Red Cross). 2010. *Bringing Help, Bringing Hope: The American Red Cross Responses to Hurricane Katrina, Rita and Wilma*. Washington, DC.

Barnshaw, J. 2005. Continuing Significance of Race and Class Among Houston Hurricane Katrina Evacuees. *Natural Hazards Observer* 30 (2): 11–13.

Bigg, M. 2010. Aid Group Shocked over Lack of Shelter for Haitians'. *Reuters*. Mar 5. http://www.reuters.com/article/2010/03/05/idUSNo5225127_CH_.2400. Last Accessed 21 May 2017.

Breennan, R.J., and R.J. Waldman. 2006. The South Asian Earthquake Six Months Later—An Ongoing Crisis. *The New England Journal of Medicine* 354: 1769–1771.

Brown, G. 2005. Making a Tragedy Out of a Disaster. *The World Today* 61 (2): 7–8.

Butterbaugh, L. 2005. Why did Huricane Katrina Hit Women So Hard? *Off Our Backs* 35 (9/10): 17–19.

Cavallo, E.A., A. Powell, and O. Becerra. 2010. Estimating the Direct Economic. Damage of the Earthquake in Haiti. IDB Working Paper Series No. IBD-WP-163, Inter-American Development Bank, Washington, DC.

Chia, E.S. 2007. Engineering Disaster Relief. *IEEE Technology and Society Magazine* 26 (3): 24–29.

Clark, N. 2005. Disaster and Generosity. *Geographical Journal* 171 (4): 384–386.

Coles, J.B., J. Zhuang, and J. Yates. 2012. Case Study in Disaster Relief: A Descriptive Analysis of Agency Partnerships in the Aftermath of the January 12, 2010 Haitian Earthquake. *Socio-Economic Planning Sciences* 46: 67–77.

Cook, J., and N. Robins-Early. 2017. Aid Workers in 4 Countries Facing Famine Warn Trump's Cut Could Cost Lives. *The Huffington Post*, Apr 12. http://www.huffingtonpost.com/entry/aid-workers-somalia-south-sudan-yemen-nigeria-famine_us_58dea89fe4b0c777f7872e18.

Cooper, C., and R. Block. 2007. *Disaster: Hurricane Katrina and the Failure of Homeland Security*. New York: Henry Holt and Company LLC.

Cruz, A.M., and E. Krausmann. 2008. Damage to Offshore Oil and Gas Facilities Following Hurricanes Katrina and Rita: An Overview. *Journal of Loss Prevention in the Process Industries* 21: 620–626.

Defense Watch. 2005. Where Was the Army National Guard When They Were Needed at Home? 20 Sept.

DesRoches, R., M. Comerio, M. Eberhard, W. Mooney, and G.J. Rix. 2011. Overview of the 2010 Haiti Earthquake. *Earthquake Spectra* 27 (S1): S1–S21.

de Ville de Goyet, C., and L. Moriniere. 2006. *The Role of Needs Assessment in the Tsunami Response*. London: Tsunami Evaluation Coalition.

Edmond, K. n.d. NGOs and the Business of Poverty in Haiti. NACLA. https://nacla.org/news/ngos-and-business-poverty-haiti. Last Accessed 11 June 2017.

Eikenberry, A.M., V. Arroyave, and T. Cooper. 2007. Administrative Failure and the International NGO Response to Hurricane Katrina. *Public Administration Review* 67 (S1): 160–170.

Elder, K., S. Xirasagar, N. Miller, S.A. Bowen, S. Glover, and C. Piper. 2007. African Americans' Decisions Not to Evacuate New Orleans Before Hurricane Katrina: A Qualitative Study. *American Journal of Public Health* 97 (S1): S124–S129.

Elliot, J.R., and J. Pais. 2006. Race, Class, and Hurricane Katrina: Social Differences in Human Response to Disaster. *Social Science Research* 35 (2): 295–321.

Financial Tracking Service. 2017. https://fts.unocha.org/appeals/overview/2017. Last Accessed 23 Mar 2017.

Flint, M., and H. Goyder. 2006. *Funding the Tsunami Response*. London: Tsunami Evaluation Coalition.

Griffin, T., M. Joseph, B. Concannon, and S.M.R. Dougherty. 2010. *Neglect in the Encampments: Haiti's Second-Wave Humanitarian Disaster*. Boston: Institute for Justice and Democracy.

GWWC (Giving What We Can). 2014. Emergency Aid, July. https://www.givingwhatwecan.org/research/other-causes/emergency-aid/. Last Accessed 16 June 2017.

Harrington, L., J. Gordon, and B.K. Paul. 2006. Southeastern Louisiana Evacuation/Non-evacuation for Hurricane Katrina. In *Learning from Catastrophe: Quick Response Research in the Wake of Hurricane Katrina*, 327–352. Boulder: Hazards Research Center, University of Colorado Boulder.

Hebblethwaite, B. 2011. Haiti Earthquake (2010). In *Encyclopedia of Disaster Relief*, ed. K.B. Penuel and M.M. Statler, 258–262. Newbury Park: Sage.

HHI (Harvard Humanitarian Initiative). 2011. *Disaster Relief 2.0: The Future of Information Sharing in Humanitarian Emergencies*. Washington, DC and Berkshire, UK: UN Foundation & Vodafone Foundation Technology Partnership.

Holguin-Veras, J., M. Jaller, and T. Wachtendrof. 2012. Comparative Performance of Alternative Humanitarian Logistic Structures after the Port-au-Prince Earthquake: ACEs, PIEs, and CANs. *Transport Research Part A* 46: 1623–1640.

Holguin-Veras, J., M. Jaller, L.N.V. Wassenhove, N. Perez, and T. Wachtendorf. 2014. Material Convergence: Important and Understudied Disaster Phenomenon. *Natural Hazards Review* 15(1): 1–12.

HRC (Hazards Research Center) (ed.). 2006. *Learning from Catastrophe: Quick Response Research in the Wake of Hurricane Katrina*, 327–352. Boulder: HRC, University of Colorado Boulder.

Iaccino, L. 2017. South Sudan Famine: Russia and US Disagree on Where Blame Lies. *International Business Times*, Mar 24. www.ibtimes.co.uk/south-sudan-famine-russia-us-disagree-where-blame-lies-1613505. Accessed 2 June.

IFRC (International Federation of Red Cross) and Red Crescent Society (RCS). 2006. Pakistan: Earthquake. *Operations Update*, 30 May.

IRIN (Integrated Regional Information Network) Pakistan. 2006. When Disasters Strike: The Response to the South Asian Earthquake, the Role of Military in the Pakistan Earthquake. www.irinnews.org/InDepthId=6&ReportsId=34370. Last Accessed 9 Jan 2007.

IRIN (Integrated Regional Information Network). 2010. Pakistan: IRIN—Critiques of the 2005 Quake Reconstruction. UNISDR, 13 May. http://www.preventionweb.net/news/view/13912. Accessed 5 June 2017.

Jenson, J.M. 2005. Reflections on Natural Disasters and Traumatic Events. *Social Work Research* 29 (4): 195–197.

Jonkman, S.N., B. Maaskant, E. Boyd, and L.M.L. Levitan. 2009. Loss of Life Caused by the Flooding of New Orleans After Hurricane Katrina: Analysis of the Relationship Between Flood Characteristics and Mortality. *Risk Analysis* 29 (5): 676–698.

Karan, P.P. 2011. Introduction: When Nature Turns Savage. In *The Indian Ocean Tsunami: The Global Response to a Natural Disaster*, ed. P.P. Karan and S.P. Subbiah, 1–31. Lexington: The University Press of Kentucky.

Kasfir, N. 2005. Sudan's Darfur: Is It Genocide? *Current History* 104: 195–202.

Kirsch, T.D., E. Leidman, W. Weiss, and S. Doocy. 2012. The Impact of the Earthquake and Humanitarian Assistance on Household Economies and Livelihoods of Earthquake-Affected Populations in Haiti. *American Journal of Disaster Medicine* 7 (2): 85–94.

Klarreich, K., and L. Polman. 2012. The NGO Republic of Haiti: How the International Relief Effort After the 2010 Earthquake Excluded Haitians from Their Own Recovery. *The Nation*, Oct 31. https://www.thenation.com/article/ngo-republic-haiti/. Last Accessed 11 June 2017.

Knox, R. 2015. 5 Years After Haiti's Earthquake, Where Did the $13.5 Billion Go? http://www.npr.org/sections/goatsandsoda/2015/01/12/376138864/5-years-after-haiti-s-earthquake-why-aren-t-things-better. Last Accessed 29 May 2017.

Kweifio-Okai, C. 2014. Where Did the Indian Ocean Tsunami Aid Money Go? *The Guardian*, Dec 25.

Large, T. 2005. Analysis—Big Tsunami Donors Rank Poorly in Generosity. *Reuters*, June 24. http://www.alertnet.org/thenews/newsdesk/L23032450.htm. Accessed 30 June 2005.

Lavelle, K., and J. Feagin. 2006. Hurricane Katrina: The Race and Class Debate. *Monthly Review* 58 (3): 52–66.

Maestas, C.D., L.R. Atkeson, T. Croom, and L.A. Bryant. 2008. Shifting the Blame: Federalism, Media, and Public Assignment of Blame Following Hurricane Katrina. *The Journal of Federalism* 38 (4): 609–632.

Margesson, R. 2005. *Indian Ocean Earthquake and Tsunami: Humanitarian Assistance and Relief Operations: CRS Report for Congress*. Washington, DC: Congressional Research Service.

Margesson, R., and M. Taft-Morales. 2010. *Haiti Earthquake Crisis and Response*. Washington, DC: Congressional Research Service.

Marshall, L. 2005. Were Women Raped in New Orleans? Addressing Human Rights of Women in Times of Crisis. *Off Our Backs* 35 (9): 14–15.

Maru, M.T. 2013. The Real Reasons Behind South Sudan Crisis: The Escalating Proves That Secession Is Not a Panacea for Peace and Democracy. *Aljazeera*, 27 Dec.

Mattera, P. 2005. Debunking Wal-Mart's Hurricane Relief Efforts. Corporate Research E-Letter No. 55. https://www.organicconsumers.org/old_articles/btc/katrina102805.php. Last Accessed 23 July 2007.

McGirk, J. 2005. Kashmir: The Politics of an Earthquake. Open Democracy, 19 Oct. https://www.opendemocracy.net/conflict-india_pakistan/jihadi_2941.jsp. Last Accessed 5 June 2017.

McGreal, C., and E. Addley. 2010. Haiti Aid Agencies Warn: Chaotic and Confusing Relief Effort Is Costing Lives. *The Guardian*, Jan 18. https://www.theguardian.com/world/2010/jan/18/haiti-aid-distribution-confusion-warning. Last Accessed 12 June 2017.

Museveni, Y. 2014. Uganda Troops Fighting South Sudan Rebels. *BBC News*, Jan 16.

News24. 2016. South Sudan Is Dying, and Nobody Is Counting. 11 Mar 2016.

Nijman, J., P.O. Muller, and H.J. de Blij. 2016. *World Today: Concepts and Regions in Geography*. Hobeken, NJ: Wiley.

Otterman, S. 2005. What Is the Controversy Over the U.N. Oil-for-Food Program? Washington, DC: Council on Foreign Relations.

Oxfam. 2011. From Relief to Recovery Supporting Good Governance in Post-earthquake Haiti. 142 Oxfam Briefing Paper. http://www.oxfam.de/sites/www.oxfam.de/files/20110106_fromreliefrecover_379kb.pdf. Accessed 25 July 2017.

Ozerdem, A. 2006. The Mountain Tsunami: Afterthoughts on the Kashmir Earthquake. *Third World Quarterly* 27: 397–419.

Paul, B.K. 2007. 2004 Tsunami Relief Efforts: An Overview. *Asian Profile* 35 (5): 467–478.

Paul, B.K. 2011. *Environmental Hazards and Disasters: Contexts, Perspectives and Management*. Hoboken, NJ: Wiley-Blackwell.

Paul, B.K., and M. Nadiruzzaman. 2013. Religious Interpretations for the Causes of the 2004 Indian Ocean Tsunami. *Asian Profile* 41 (1): 67–77.

Paulsen, D. 2017. Episcopal Agencies Weigh Renewed Aid Efforts in South Sudan After Famine Declared. *ENS News Service*, Feb 29.

Pew Research Center. 2005. *Two-in-Three Critical of Bush's Relief Efforts: Huge Racial Divide Over Katrina and Its Consequences*. Washington, DC.

Pew Research Center. 2010. Haiti Dominates Public's Consciousness: Nearly Half Have Donated or Plan to Give. Washington, DC: Pew Research Center—20 Jan. http://people-press.org/2010/01/20/haiti-dominates-publics-consciousness/. Last Accessed 29 May 2017.

Quinn, B. 2017a. Famine Declared in South Sudan. *The Guardian*, Feb 20.

Quinn, B. 2017b. UK's £100m Response to South Sudan Famine Comes from Cash Already Allocated. *The Guardian*, Feb 23.

Quinn, B. 2017c. EU and UK United in Effort to Combat Famine in South Sudan. *The Guardian*, Feb 22.

Qureshi, J.H. 2006. Earthquake Jihad: The Role of Jihadis and Islamist Groups After the 2005 Earthquake. HPN, July. http://odihpn.org/magazine/earthquake-jihad-the-role-of-jihadis-and-islamist-groups-after-the-2005-earthquake/. Last Accessed 5 June 2017.

Rashid, A. 2005. Post-quake Giving United Pakistan. *The New York Times*, Oct 27.

Real World Survivor. 2015. 5 Deadliest Natural Disasters Since 2000, 23 Feb. http://www.realworldsurvivor.com/2015/02/23/5-deadliest-natural-disasters-since-2000/#bularnw-nl. Last Accessed 2 June 2017.

Renner, M., and Z. Chafe. 2007. *Beyond Disasters: Creating Opportunities for Peace*. Washington, DC: Worldwatch Institute.

Renton, A. 2005. Money, Not Excuse. *The Guardian*, Nov 20.

Rofi, A., S. Doocy, and C. Robinson. 2006. Tsunami Mortality and Displacement in Aceh Province Indonesia. *Disasters* 30 (3): 340–350.

Schuller, M. 2010. Haiti's Disaster After the Disaster: The IDP Camps and Cholera. *Journal of Humanitarian Assistance* (13 Dec).

Seto, T., and Y. Nishimura. 2016. Crisis Mapping Project and Counter-Mapping by Neo-Geographers. In *Japan After 3/11: Global Perspectives on the Earthquake, Tsunami, and Fukushima Meltdown*, ed. P.P. Karan, and U. Suganuma, 288–303. Lexington: University Press of Kentucky.

Sieff, K. 2017. South Sudan's People Are Starving, and Fighters Are Blocking Aid. *The Washington Post*, Mar 31.

Slagh, C.L. 2010. Managing Chaos, 140 Characters at a Time: How the Usage of Social Media in the 2010 Haiti Crisis Enhanced Disaster Relief. A Master thesis, Georgetown University, Washington, DC.

Spiegel, P.B. 2005. Differences in World Response to Natural Disasters and Complex Emergencies. *Journal of American Medical Association* 293 (15): 1915–1918.

Storm, S. 2006. Foreign Experts Critique U.S. Red Cross on Katrina. *The New York Times*, Apr 5.

Stout, D. 2005. Bush and Congress Announce Inquiries on Federal Response. *The New York Times*, Sept 6.

Street, A., and G. Parihar. 2007. The UN Cluster Approach in Pakistan Earthquake Response: An NGO Perspective. http://odihpn.org/magazine/the-un-cluster-approach-in-the-pakistan-earthquake-response-an-ngo-perspective/. Accessed 21 May 2017.

Tamamoto, M. 2005. After the Tsunami, How Japan Can Lead. *Far Eastern Economic Review* 168 (2): 15–19.

Telford, J., and J. Cosgrave. 2007. The International Humanitarian System and the 2004 Indian Ocean Earthquake and Tsunamis. *Disasters* 31 (1): 1–28.

The Economist. 2006. Pakistan's Earthquake. 6 May.

The Henry J. Kaiser Family Foundation. 2005. *Survey of Hurricane Katrina Evacuees*. Washington, DC: The Henry J. Kaisser Family Foundation.

Thomas, E. 2005. What Went Woring: The Story of Katrina—And a Disastrously Slow Rescue. News Week (US Edition), Nov 17, http://www.newsweek.com/what-went-wrong-story-katrina-and-disastrously-slow-rescue-118069. Last Accessed 16 Nov 2017.

Travis, J. 2005. Scientists' Fears Come True as Hurricane Floods New Orleans. *Science* 309: 1656–1659.

Ugra, S. 2005. Island of Apathy. *India Today*, Jan 17, 8–12.

U.N. (United Nations). 2006. *Tsunami Recovery: Taking Stock After 12 Months*. New York.

U.N.-ESC (United Nations-Economic and Social Council). 2005. *Strengthening Emergency Relief. Rehabilitation, Reconstruction, Recovery and Prevention in the Aftermath of the Indian Ocean Tsunami Disaster*. New York.

US Department of Health and Human Services. 2010. Information for U.S. Citizens Returning from Haiti. http://www.hhs.gov/haiti/repatriationprogram.html. Accessed 26 Feb 2010.

US Department of Homeland Security. 2010. Statement from Homeland Security Secretary Janet Napolitano on Temporary Protected Status (TPS) for Haitian Nationals. Jan 15 2010. http://www.dhs.gov/ynews/releases/pr_1263595952516.shtm. Accessed 4 Feb 2010.

US Institute of Peace. 2010. Haiti: A Republic of NGOs? Peace Brief 23, 26 Apr. https://www.usip.org/sites/default/files/PB%2023%20Haiti%20a%20Republic%20of%20NGOs.pdf. Last Accessed 11 June 2017.

US State Department. 2013. Foreign Aid to U.S. for Katrina Relief. http://www.nbcnews.com/id/9282598/ns/us_news-katrina_the_long_road_back/t/foreign-aid-us-katrina-relief/#.Wg4E4VVKthE. Last Accessed 16 Nov 2017.

Valley, P. 2006. Review of the Year: Disasters—Natur's Assaults Have Shaped a New Reality for Mankind. *The Independent*, March 13, 11(4).

Versluis, A. 2014. Formal and Informal Material Aid Following the 2010 Haiti Earthquake as Reported by Camp Dwellers. *Disasters* 39 (S1): S94–S109.

Waugh Jr., W.L. 2005. The Disaster That Was Katrina. *Natural Hazards Observer* 30 (7): 7–8.

Wehrfriz, G., and J. Cochran. 2005. Charity and Chaos. *Newsweek* 17: 30–32.

Wilhelm, I. 2005. Sharing a World of Experience: Foreign-Aid Groups Offer Advice in Hurricanes' Aftermath. *Chronicle of Philanthropy*, Oct 6. http://philanthropy.com/free/articles/v18/i01/01002701.htm. Accessed 6 Aug 2006.

Zimmermann, K.A. 2015. Hurricane Katrina: Facts, Damage & Aftermath. *Life Science.* 27 Aug. http://www.livescience.com/22522-hurricane-katrina-facts.html. Last Accessed 3 June 2017.

5
Convergence Phenomenon

Donating unnecessary and useless emergency relief items to those suffering either a natural or a man-made disaster has been a perpetual problem across the globe. The spontaneous flow of volunteers, emergent and traditional relief organizations, information, and supplies to disaster and emergency sites from places far from the disaster often hampers and complicates emergency relief efforts (Nelan 2016). This post-disaster convergence phenomenon has three forms: personnel (the influx of public emergency response personnel and all types of individuals, including affiliated or non-affiliated local/national and foreign volunteers, members of search, rescue, and medical teams, and personnel associated with humanitarian and other relief agencies); informational (the flow of information and communication); and material (movement of physical resources like (solicited or unsolicited) donated goods, supplies, and/or equipment sent by all entities in response to a disaster) (Fritz and Mathewson 1957; Cole-Corde 2013; Holguin-Veras et al. 2014a). These components of convergence and their effects on disaster relief activities are examined in this chapter.

B. K. Paul, *Disaster Relief Aid*,
https://doi.org/10.1007/978-3-319-77282-0_5

Convergence is the result of a behavioral response of good-hearted people and a host of organizations not directly affected by environmental disasters, but who donate, collect, or provide support to those affected by extreme events.[1] Prince (1920) first identified the post-disaster convergence phenomenon in his doctoral research on the Halifax, Canada, ship explosion. Since then this phenomenon has been reported in almost all major disasters, including 2005 Hurricane Katrina, 2012 Hurricane Sandy, the 2010 Haiti Earthquake, and the 2011 Japanese Tsunami. Fritz and Mathewson (1957) defined convergence as the movement of solicited and unsolicited donated goods and other items sent by donors of all kinds toward disaster and emergency sites and locations associated with the event. Despite its relatively long history, the phenomenon of disaster convergence deserves more attention from hazard researchers because the phenomenon continues to repeat itself in disaster after disaster. More research could help alleviate the challenges of convergence for disaster supply chains as well as survivors of extreme events.

In this chapter, we will discuss three types of post-disaster convergence, the advantages and disadvantages of the convergence phenomenon, and strategies to control convergence. The chapter will conclude with recommendations for stopping the flow of unnecessary relief items, personnel, and information from unaffected areas to disaster-affected areas.

Convergence of Personnel

Fritz and Mathewson (1957) identified five kinds of personal who converge on a disaster area. The first are returnees, people who reside in the disaster-affected area but evacuated the area before the extreme event occurred. They may have left the area because of a curfew was imposed after the event. The second type are the anxious, people from non-affected areas who converge on the affected site to obtain information about family, relatives, and friends. The third type are the people who also converge at the site to provide emergency help to disaster survivors; they are called helpers by Fritz and Mathewson (1957). The fourth are curious, those who come to the affected site to see the destruction caused by the disaster. Exploiters are the fifth type, who come to the affected area for personal gain, generally before curfew is imposed and law enforcement authorities arrive. This type of convergence is directly associated with one of the myths of a disaster: looting.[2]

Kendra and Wachtendorf (2004) added a sixth and a seventh type of personnel convergence: mourners and fans or supporters (also see Amanda et al. 2008). Supporters usually gather at a disaster site to express gratitude to emergency workers and volunteers. This happened after the World Trade Center attack in New York and many other disasters. In this chapter, convergence of personnel in the immediate aftermath of a disaster refers to people who directly participate in search and rescue (SAR) operations, provide emergency medical care, and disburse emergency assistance among the survivors. Fritz and Mathewson (1957) called them helpers.

SAR operations include locating victims who are injured and/or trapped under collapsed buildings or rubbles, rescuing them as quickly as possible, providing them basic first aid, and if needed, transporting them to nearby medical facilities for further treatment (Paul 2011). Most initial disaster SAR is carried out by family members, neighbors, relatives, coworkers, and others who happen to be in the affected area. Later, they are joined by nearby trained and professional rescue workers. About half of all rescues occur in the first 6 hours after a disaster. Trained SAR teams, who usually come from within the disaster-affected country, but outside the disaster-affected areas rescue the other half after they arrive (Paul 2011).

In large disasters, rescue workers, medical teams, and volunteers from abroad also converge at disaster sites, as was the case after the 2003 Iranian Earthquake, 2004 Indian Ocean Tsunami (IOT), 2010 Haiti Earthquake, and 2015 Nepal Earthquakes. Very few developing countries have the expertise to conduct SAR operations, so SAR teams are often dispatched by industrial nations. However, SAR has rapidly diminishing returns. Those who are trapped often die very quickly without treatment. SAR teams from foreign countries take too long to reach the disaster zone, rendering their high cost effectively wasted.

Some 1600 foreign rescuers arrived 36 hours after the quake in the city of Bam, Iran, the center of the devastation. None of the rescuers reached the earthquake-affected area during the critical six to eight hours after an earthquake when urban search and rescue (USAR) are most needed. These foreign rescuers found only 30 people alive in the debris in Bam (Abbasi and Salehnia 2013). The cost of such interventions averages around one million dollars per life saved. If local SAR teams were deployed, that cost could be reduced to 50 cents per life saved. Moreover, many logistical and communications problem could be avoided (PAHO 2003). In fact, Abbasi and Salehnia (2013) reported that the Iran Red Crescent Society had saved the lives of 1000 people in the first 24 hours.

After the 2004 IOT, many rescue teams came to Thailand from Asia, Europe, and the USA. Paul (2011) reported that nearly 2000 rescuers with 161 search dogs from Argentina, Britain, Canada, China, the Dominican Republic, Iceland, the USA, and other countries converged on Haiti two days after the country was hit by a 7.0 magnitude earthquake in 2010. Members of these foreign SAR teams, particularly those from the USA, were well equipped and thus saved many lives of people who were trapped in collapsed buildings.

Assuming that the 2015 Nepal Earthquakes was an urban disaster, 23 countries sent 53 USAR teams into Kathmandu, including 1872 personnel and 177 search dogs. In addition, at least 80 foreign medical teams registered to work in Nepal to treat people injured by earthquakes. The rescue teams were able to rescue only 16 people. Most of the deaths caused by the earthquake occurred in remote rural areas. Because of the remoteness and hazardous road conditions, teams could not reach the severely affected rural areas (Troutman 2015). Roads were severely damaged by earthquakes, and the many aftershocks and landslides made it nearly impossible to reach already isolated mountain villages (Paul et al. 2017). The teams did use available planes and helicopters to find and treat those injured in the earthquake. However, the Nepalese government was overwhelmed with the logistical demands of the teams, and three days after the quake, the government asked foreign medical and SAR teams to leave the country (Zakaria 2015). Troutman (2015) reported that an additional 1309 foreign medical and 152 SAR personnel arrived after the government's plaintive prohibition. Upon arrival, many found little to do because local Nepalese were already doing a superb job of digging out as many bodies as possible (Zakaria 2015).

Troutman (2015) further reported that beginning on May 1, 2015, the SAR teams were directed to recover dead bodies, work that did not require the level of expertise foreign rescuers have. On average, 117 foreign rescuers rescued only one person. On May 5, the Nepalese government finally asked medical, search, and rescue teams to leave because they expected the government to provide them food, water, transportation, and other support. Providing logistics to foreign teams was one barrier confronting the public authorities of Nepal in offering timely and necessary services to earthquake survivors. In fact, nine foreign medical teams came to Nepal and left without talking to anyone (Troutman 2015).

Convergence of a plethora of private, public, humanitarian, and nonprofit organizations, as well as individuals from both affected and non-affected areas/countries during the emergency-relief phase after a major disaster, particularly in developed countries, has now become the norm. As noted in Chapter 2, national and foreign armed forces also often respond to disaster. In recent decades, the 2004 IOT probably set the record for convergence of personnel. In the first week after the 2004 IOT, many teams from the U.N. and other multilateral organizations, high-profile government personnel, and hundreds and thousands of individuals both affiliated with relief organizations and on their own, from both affected and non-affected countries participated in disaster relief efforts. Chia (2007) reported that in the aftermath of the tsunami, approximately 700 NGOs from 40 different countries responded.

Apart from problems with coordination, convergence of volunteers from abroad and non-affected areas creates many logistical problems. They can create congestion and often actually interfere with relief operations (Tierney 2003). For example, most foreign volunteers arrived in the 2004 tsunami-affected countries, particularly in India, Indonesia, and Sri Lanka, without their own resources or appropriate training to assist tsunami survivors (U.N. 2005). In most cases, their participation did not facilitate relief operations and instead delayed relief distribution (Paul 2007). Delays in relief distribution may also be caused by convergence of emergency response volunteers coming from outside the affected areas because they are not familiar with the areas and may not have adequate knowledge of local norms. Local and national authorities responsible for relief operations must also divert their attention and valuable time from relief efforts to provide food, accommodation, and transport for these volunteers. A U.N. report rightly called this convergence of personnel a "humanitarian traffic jam" (UN–ESC 2005).

As Miller (2005) noted, it was nearly impossible to keep track of foreign relief personnel coming into Sri Lanka and what they were doing. He claims that this huge influx of individuals was partly responsible for the failure of Sri Lankan government in distributing tsunami relief goods. Miller (2005) also reported that more temporary shelters were provided for these foreigners involved in relief operations in Sri Lanka than for tsunami survivors themselves. In Thailand, "There were more NGOs looking for villages to work in than villages available" (Plan International 2005).

So many organizations and individuals are involved in responding to a disaster that responders absolutely must coordinate, communicate, assess the ongoing situation, and mobilize resources during the emergency period. Coordination is generally easier if fewer agencies are involved in disaster relief operations. More agencies means more time for negotiating, sharing, and coordinating than providing for the needs of disaster survivors (Paul 2007). Agencies also compete each other for space. In India, between 500 and 600 organizations were believed to be involved with tsunami relief distribution by the end of January 2005. This led to a situation where there were too few accommodations for the personnel of these agencies (Plan International 2005). In all types of disasters, communication is needed among policy-makers, first responders, volunteers, and emergency managers to work effectively and successfully during the emergency.

Lack of communication and coordination among participating agencies often results in wasting a sizable share of the resources flowing into the disaster-affected area, which also prolongs the suffering of disaster survivors and delays recovery efforts (Yonder 2005). Thus, lack of coordination often becomes the primary reason for the failure of a response to a disaster. At the global level, the United Nations Office for the Coordination of Humanitarian Affairs (OCHA) typically coordinates humanitarian emergency responses by establishing an appropriate response mechanism and facilitating smooth distribution of relief and other emergency needs to disaster victims. However, many agencies ignore OCHA. For example, by the third week of January 2006, there were at least 200 agencies on the ground in Aceh, Indonesia, but less than a quarter of these agencies had contacted OCHA (IFRC 2005). As a result, duplication of relief efforts has been widely reported in several tsunami-affected countries, but especially in India and Sri Lanka (Trainor and Kendra 2005).

Other major natural disasters have occurred since the 2004 IOT, among them the 2010 Haiti Earthquake and 2015 Nepal Earthquakes, have also experienced convergence of personnel and organizations. The convergence phenomenon is also a problem for complex emergencies. For example, convergence of spontaneous volunteers and emergent groups after the September 11, 2001, terrorist attack in New York City proved more problematic than beneficial in the overall response to this event. Most of these volunteers were domestic, had little training or experience, and represented people from all walks of life and many professions. They created many logistical problems for public responders. Volunteer firefighters gathered along the West Side Highway which caused massive confusion and blocked critical access for getting resources to and from Ground Zero (McKay, n.d.).

However, volunteers or helpers both from within and outside the disaster-affected countries and emergent groups are necessary for successful relief operations. These volunteers can meet legitimate needs at disaster sites (Kendra and Wachtendorf 2004). However, too few or too many volunteers make relief operations unsuccessful. In the case of too many volunteers and agencies at work at a disaster site, efficient management and coordination of the volunteers are needed to avoid a "second-tier disaster" (McKay, n.d.). Currently, no one has suggested an ideal size for this flow of volunteers. Disaster researchers, managers, and humanitarian organizations must determine how many volunteers are needed based on magnitude of the event and the resources of the affected country.

Given this convergence of people as a universal post-disaster phenomenon, all volunteers must be educated about how their physical presence in disaster sites is potentially disruptive. Additionally, to properly manage the volunteers, they must be properly documented to verify their identities and associations with relief organizations, if any. The relevant authorities must inquire whether they have the training, experience, and skills to perform tasks in disaster locations. One way to control the number of foreign volunteers is for the authorities of disaster-affected countries to impose a quota system.

Convergence of Information

Informational convergence is the influx of information into disaster-affected areas, generally caused as both traditional and modern media gather and/or repeatedly disseminate information about an event, which often makes a small disaster seem large. This type of convergence coincides with personal convergence. Among all three convergences, this one is among the least studied, and little material is available on this type of convergence. Convergence of information has both advantages and disadvantages. For major disasters, traditionally both electronic and print media from outside the affected areas converge at the disaster site and play a critical role in structuring the relief response by framing emergency supply needs and rapidly disseminating information on those needs throughout affected and non-affected countries. Both domestic and foreign electronic media can mobilize the relief process.

The number and length of both national and international media reports often determine what emergency relief flows to the disaster-affected areas. Coverage can lead to a flood of inappropriate donations. For example, the mass media covering Hurricane Katrina in 2005 initially were unaware of several heavily impacted wards in New Orleans. For this reason, survivors in these wards received no emergency aid. Later, when the media discovered and broadcast the situation in these wards, relief items started to flow (Wachtendorf et al. 2010). A similar situation was also reported in areas of several countries affected by the 2004 IOT and in the aftermath of the 2008 earthquake in China (Wachtendorf et al. 2006; Letukas 2008; Wachtendorf 2008).[3]

Holguin-Veras (2012) claimed that several days after 9/11 terrorist attacks on the World Trade Center, a TV news broadcast that the feet of SAR dogs were getting burned. This prompted pet stores, dog shoe manufacturers, and animal lovers to send huge quantity of dog shoes and dog food. These donations were not useful because rescue dogs have specific dietary needs and the dog shoes were not appropriate for rescue dogs to hunting individuals in the rubble (Wachtendorf et al. 2015). Hologuin-Veras (2012) further reported that media announced a shortage of extension cords and within a day or so, "huge numbers of them started to arrive, overwhelming both the storage capacities and actual needs on the ground" (Holguin-Veras 2012). A similar reaction was observed in a community affected by the 2013 tornadoes in Oklahoma. Nelan (2016) reported that because a tornado-affected community needed work gloves and a local news station requested donations of gloves, the community received more gloves than they needed.

However, often the media identify the needs of disaster survivors before government emergency agencies do so. Thus, the media can serve to push the donation and supply chain to particular locations within a disaster-affected area.

As indicated in Chapter 1, social media like Facebook, Twitter, YouTube, and other modern information sharing platforms, including crowdsourcing, has been widely used since the 2010 Haiti Earthquake.[4] During the response phase of disasters, emergency responders and relief agencies can use social media information for timely rescue of people trapped in the rubble and to disburse disaster relief to afflicted people more quickly. Because of social media, both medical and public health professionals are also better able to respond to disasters (Merchant et al. 2011). Further, this modern technology can help track population movements during disasters, making distribution of emergency aid more effective (*Natural Hazards Observer* 2011). Benglsson et al. (2011) analyzed position data from 1.9 million SIM cards from mobile phones in Haiti after the 2010 earthquake. They found that estimates of population movement derived from SIM card location were more accurate than ad hoc estimates. In Syria, digital data collection tools have been using over the last several years to track commodity movement and monitor the distribution of relief goods by local partner organizations in areas that are inaccessible to international humanitarian agencies (IFRC 2013).

Using two datasets of individual mobility trajectories from six million mobile phone users in Bangladesh, Lu et al. (2016), after removing identifying information, examined the mobile network data to explore important short-term movement (hours-weeks) during and after Cyclone Mahasen, which struck Bangladesh in May 2013. They showed that mobile data allowed them to study the relationship between fundamental parameters of migration patterns on a national scale. They also quantified incidence, direction, duration, and seasonality of migration in Bangladesh. Thus, social media can serve as efficient communication platforms at individual and other levels to exchange emergency information quickly.[5] During the 2011 Japan Earthquake and Tsunami, as well as during New Zealand's Christchurch Earthquake in 2010, crowd-sourced, crisis mapping sites provided by the *Ushahidi* and Google maps proved crucial in helping local people and relief organizations identify communities that urgently needed relief.[6] Because so many people use social and other media, however, potential convergence of these sources in disaster sites can occur, as was the case in 2010 in Haiti and in 2015 in Nepal.

However, not all types of media have complete and accurate information on the needs of survivors of a particular disaster (Wachtendorf et al. 2010). Media, particularly social media, often hinder relief operations by spreading false or misleading information (Alexander 2014), or falsely notifying officials of a disaster or threat that requires a response (Lindsay 2011). This problem, to a considerable extent, results from lack of time to verify the accuracy of information received. A report published in *Natural Hazards Observer* (2011) noted that a brief and revealing social media message sent immediately prior to the IOT might have lured people to rush to the coast. The message informed the people that sea water had receded unprecedented distances from the beach and the shell collecting was excellent. Although this cannot be verified, disaster experts and managers worry about this use of social media.

Similarly, Brenhouse (2011) reported that after the 2011 Fukushima nuclear meltdown in Japan, social media in China claimed that iodized salt can prevent radiation-related illness. As a result, large numbers of people went to grocery stores to buy and hoard salt.

This not only created chaos, but led to an unprecedented salt shortage across China. False news and rumors also spread through social media, as reported immediately after the 2004 IOT, 2010 earthquake in Chile, and Hurricane Sandy in 2012 in the USA (Alexander 2014). Social media companies need to develop devices that effectively control rumors.

After the 2015 Nepal Earthquakes, conventional and social media informed the public that during future earthquakes or aftershocks, they should take shelter under beds inside homes. Accordingly, during many aftershock events, people from outside took shelter as directed. This misinformation provided by social media to earthquake survivors in Nepal. In fact, it is safer to stay outside during an earthquake.[7] Usually, however, misinformation is corrected by moderators or users of social media. Disaster responders and emergency managers must promptly identify inaccurate and misleading information and provide correct information to disaster survivors and others. Moreover, the public must be taught how to use social media effectively in an emergency.

The spread of technology also creates challenges for social media in disaster relief operations. The amount of access to technologies differs among both affected populations and humanitarians. The poor, the illiterate, and women, particularly in developing countries, have little or no access to the technology. Similarly, the first responders to disasters, the local humanitarian organizations, and local governments are also least able to afford access to modern technology. One structural limitation is the expensive communication bandwidth needed to improve public–private partnerships so that mobile phone network operators and internet providers can ensure minimum service (IFRC 2013).

Convergence of Materials

Material convergence is a complex problem, with multifaceted logistical challenges. Fritz and Mathewson (1957) defined material convergence as the movement of solicited and unsolicited donated goods, supplies, and other items sent by donors of all kinds toward disaster sites and other locations affected by a disaster. In 1957, Fritz and Mathewson identified six characteristics of material convergence. These characteristics are still relevant today (see Holguin-Veras et al. 2014a):

- Exceed relief needs;
- Comprise materials that are both unnecessary and unusable;
- Require personnel and facilities needed elsewhere;
- Cause conflict among agencies or the people needing relief supplies;
- Cause congestion at the disaster site; and
- Disrupt the local economy.

Material convergence is often dubbed "a second-tier disaster" or simply "second disaster" (France and Ordonez 2002; Wachtrndorf et al. 2010; Holguin-Veras 2012; Nelan 2016). Material convergence occurs as two types: formal and informal. Formal material convergence is the movement of solicited donations from public–private-sector sources to people authorized to handle disaster relief operations, essentially, the transfer of solicited donated goods, as pre-established through legally binding memoranda of understanding, mutual aid agreements, or emergency management assistance compacts. Transfer of relief goods usually occurs after an official emergency declared.

Informal material convergence, in contrast, is the spontaneous flow of unsolicited material donations to disaster-affected areas, which are not generally useful to disaster survivors. Managing such donations creates many problems for people involved in collecting, storing, and disbursing relief goods. However, some unsolicited supplies often are needed, supplementing gaps in formal supply chains (Wachtendorf et al. 2010). Informal donors include individuals, relatives, neighbors, friends, and others, both domestic and international (Fritz and Mathewson 1957; Cole-Corde 2013).

The Pan American Health Organization (PAHO) classified donated goods as follows: urgent or high-priority (HP) items, required for immediate distribution and consumption; non-urgent or low-priority (LP) items, not immediately required but useful later on; and finally non-priority (NP) items, unnecessary items that should not have been sent to the disaster site at all (PAHO 2001). The inflow of large volumes of NP items creates a serious problem at disaster sites, as does, to a lesser extent, the flow of LP items (Holguin-Veras et al. 2012a). Beamon and Blanik (2008) designate tiers of donated items based on usefulness following a disaster. Tier 1 items are critical and include food, blankets, shelter, and hygiene supplies. Tier 2 items are not as critical as Tier 1 items immediately following the event.

For material convergence, Holguin-Veras et al. (2012a) established the term humanitarian logistics, which included "a broad spectrum of relief operations ranging from the regular distribution of critical supplies, such as distributing food to fight hunger, to the distribution of critical supplies after a disaster" (Holguin-Veras et al. 2014a, 1). The former they termed regular HL (R-HL) and the latter post-disaster HL (PD-HL) (also see Holguin-Veras et al. 2014b). This chapter focuses on PD-HL which they considered one of the least understood areas of disaster research. Globally, some 70 million people experience one or more types of disasters per year (Dewaal 2017) and need PD-HL. All actors in disaster relief know to some extent what survivors of an extreme event need urgently, but still they send LP and NP items, which often travel from one end of the world to the other end.

Existing literature on material convergence provides many examples of inappropriate, unnecessary, or redundant in-kind donations from external sources to disaster survivors. Fritz and Mathewson (1957) first formally documented such donations in the 1950s; the practice is still a prominent problem today (Wachtendorf et al. 2010; Holguin-Veras et al. 2007, 2014a; Cole-Corde 2013). What follows is a list of specific examples of unnecessary and even inappropriate donations. For the convenience of analysis, these items are presented below using four somewhat overlapping categories. Problems associated with these materials are also briefly discussed.

Inappropriate, LP, and Climatic Insensitive Items

In April 1921, Italian Senator Giovanni Ciraolo wrote: "Frequently the materials provided for purposes of relief are not adapted to the requirements of the people and the local climatic or economic conditions" (Elia

1928, 3). Unfortunately, this problem is repeated many times over. For example, a large quantity of useless, inappropriate, LP, and materials inappropriate to the climate arrived from abroad for the 2004 IOT survivors. Winter coats and heavy blankets were sent for survivors in the tsunami-affected countries, which enjoy a year-round warm tropical climate (Paul 2007). Winter coats were also distributed among the 1988 Hurricane Georges survivors in the Dominican Republic, which likewise has a tropical climate (McEntire 1999). This also occurred after Hurricane Andrew hit Florida in 1992. Neal (1994, 25) writes: "… if you took all the heavy winter coats, you could fill 25 warehouses in the Miami areas – Coats that in a lifetime would never be worn in Miami." Similarly, heavy blankets were sent to Japan after the Great East Japan Earthquake in 2011, but a week after the event, when the weather warmed up, the blankets were no longer needed (Holguin-Veras et al. 2012b). Ito et al. (2014) said these items were useful but not at that time. In addition, survivors of the 2011 Japan Earthquake also received completely useless items like high heels and dirty or used clothes (Ito et al. 2014).

In addition to unneeded heavy winter coats and blankets and out-sized, filthy, low quality, used, or inappropriate garments, a container full of teddy bears was sent from abroad to Sri Lanka for tsunami survivors. These bears were of no use to tsunami survivors (Paul 2007). Gautham (2006) provided still another example of unnecessary donations, reporting that after the 2004 IOT, thousands of packets of foreign sanitary napkins were distributed among tribal women in Nicobar Island, India, who had never used them and did not know what to do with them. After the 2010 floods in Colombia, NP items included used wedding and party gowns and used underwear (Holguin-Veras et al. 2014a). Perhaps most surprising and unnecessary was a truckload of sex toys that arrived at one of the distribution centers in Florida after the state was hit by Hurricane Charley in 2004 (Holguin-Veras et al. 2014a).

After the 2004 IOT, boats distributed among fishermen in Tamil Nadu, India, could not be used because they were not designed for commercial fishing (P'Rayan 2005). Fishermen in Thailand received money to replace their boats, but since they received nothing to replace their lost fishing nets, they could not resume work (Peng 2005).

In Indonesia, destroyed boats were replaced, but most were not expected to last beyond 18 months because they were poorly designed and built of substandard materials (Paul 2007). Similarly, Holguin-Veras and his colleagues (2014a) reported that 10 containers of European refrigerators were sent to Haiti after the 2010 earthquake, but these refrigerators were unusable not just because of the power outage but because the voltage did not match the Haitian electrical system.

After Cyclone Aila struck the south-west coast of Bangladesh and the coast of West Bengal, India, in 2009, NGOs distributed mosquito nets among the cyclone survivors in Bangladesh. These nets were washable only by machine, which were available only in large cities like Dhaka. Additionally, the NGOs distributed tents not suited for high temperatures. For this reason, the tents were damaged within a few months (Mahmud and Prowse 2012).

After the 2015 Nepal Earthquakes, some aid groups distributed reusable cloth sanitary pads among women survivors of remote villages. Unfortunately, very few knew what these pads were used for. The villagers reportedly turned them into the sorts of masks ubiquitous in the Kathmandu valley due to smog (ABE Streep 2015). Additionally, aid groups provided them "half pants" and t-shirts. These clothes were neither culturally nor environmentally friendly, particularly for women of rural Nepal who, because of high elevation and low temperatures, do not usually wear pants, particularly half pants. Also, many earthquake survivors could not use clothes distributed among them because of their size and shape. Most of the pants were too big for them (Paul et al. 2017).

Another necessary item for the homeless earthquake survivors of Nepal was tents. In response to an appeal by the Government of Nepal, however, international actors sent tents that were either very large or very small. The huge tents were of no use, while the small tents were too small for households with many family members (Daniel and Mahr 2015). These tents protected the survivors from rains and storms reasonably well during monsoon season, but failed to protect them from the winter cold. Nepalese daily newspapers reported deaths associated with hypothermia among the elderly, women, and children (*Kathmandu Post* 2015a).

Even worse, dew dropping from the roofs of tents or tarps was a serious problem for the earthquake survivors during the following winter in Nepal. The tents had very low roofs, just over the head, so water drops to survivors who slept inside the tents, making them sick. Also, the tents were small, lacking a separate room for women and girls to change their clothing: They felt very uneasy changing their clothing in such close proximity to male family members.

Most of the unneeded and non-priority relief items sent from abroad after each major natural disaster to the affected countries end up being dumped by the on roadside, often proving hazardous to local livestock who tried to eat them (see Paul 2007). Emergency response workers must then divert attention and resources away from genuine relief efforts to collect these useless materials. They waste precious time on these supplies, and as a result, disaster survivors suffer because they do not receive critically needed supplies as quickly as possible (Nagurney 2017). Dumped clothes block roads and consume storage space (IFRC 2005, 2006), and the delivery and processing of LP and NP supplies add to congestion at transportation and distribution nodes. No question but that inappropriate donations impede distributing necessary items among disaster survivors (McEntire 1999; Arnette and Zobel 2016).

These useless goods also occupy storage space that could be used to store more valuable, more necessary relief supplies. Generally, space for storing relief goods in developing countries is very limited. Apart from space, another problem with the outpouring of donations of unnecessary items occurs because volunteers must spend time sorting and disposing of these items, which is a significant drain on resources. Alexander (2000) calls the arrival of unnecessary donations heartless and cynical, a form of dumping surpluses on survivors of disaster, an easy way of avoiding proper disposal at home while appearing to be generous. Distribution of inappropriate relief items can insult disaster survivors of developing countries, which can be detrimental to their dignity (U.N. 2005).

After reviewing emergency relief materials sent to the survivors of several past natural disasters, Arnette and Zobel (2016) claim that inappropriate or low-priority relief goods often exceed 50% of the unsolicited cargo that arrives in disaster-affected countries. Nagurney (2017) claims that about 60% of the items that arrive in disaster-affected areas are low priority or unnecessary. Similarly, Holguin-Veras et al. (2014a) claim that 80% of clothing donations sent to Haiti after the 2010 earthquake were useless. In contrast, 80% of items received after the 2013 Colorado Flood were appropriate and usable (Arnette and Zobel 2016). This flood received only a fraction of the national news coverage provided other disasters such as Superstorm Sandy. Less media coverage contribute to lower levels of material convergence.[8]

Culturally and Religiously Insensitive Items

Often relief materials distributed among disaster survivors are overtly offensive from cultural and/or religious points of view. For example, after the 2004 IOT, Christmas carol songbooks were sent from Britain to Muslim tsunami survivors in the Maldives (Paul 2006). Additionally, several foreign Christian NGOs distributed Bibles among Muslim tsunami survivors, creating tensions in Indonesia and upsetting both the survivors and the government. One consequence was that the Indonesian military expelled all NGOs involved (Brown 2005).

In a similar example of cultural insensitivity, after the 1974 flood in Bangladesh, Western countries donated massive quantities of canned pork to distribute among flood-affected people, ignoring the fact that Bangladesh is a Muslim-dominated country and Muslims do not consume pork (Paul 2006). This was repeated after the 2004 IOT; meals containing pork were sent to Muslim tsunami survivors of Aceh, Indonesia (Telford and Cosgrave 2007; Bradshaw 2013). During the 2010 flood in Pakistan, relief camp personnel served beef to Hindus who were taking refuge there (Spencer 2010). Following the IOT, commercial tents were shipped to some tsunami-affected countries of Asia to provide temporary shelters for households. These tents were too small for their large families and uncomfortably warm during the afternoon hours (Wachtendorf et al. 2015). Many survivors refused these tents, constructing their own temporary shelters using tarps, thatch, and tree branches.

After the devastating earthquakes in Nepal in 2015, several Indian newspapers published reports that Pakistan had sent packets of ready-to-eat meals, including packets of "beef masala," as part of its aid package (Saxena 2015). The overwhelming majority of Nepalese are Hindus, and therefore, this gesture was considered offensive, causing outrage throughout the Nepalese media. According to Hindu belief, eating beef is a religious offense because the cow is a sacred animal. The first Civil Codes of Nepal, the Muluki Ain of 1854, prohibited slaughter of cows, with death and confiscation of all property of the offender as punishment. An amendment in 1990 to the Civil Code made cattle slaughter punishable by 12 years in prison (*Mail Today* 2015).

The Nepalese authority identified and kept aside cartons containing the controversial food items sent from Pakistan. Pakistani authorities, however, claimed they had sent no beef in the ready-to-eat food dispatched to Nepal and urged the Indian media not to "malign" humanitarian assistance effort in Nepal (*Express Tribune* 2015). Participants of several focus group meetings (FGMs) conducted by this author in January 2016 (Paul et al. 2017) included questions about receiving any beef masala packets as emergency aid. None had received any such packets although participants reported hearing about others who had.

However, other food items, such as cans of tuna and mayonnaise, sent by foreign sources to Nepalese Earthquake survivors, were either not culturally appropriate or not edible (*Telegraph* 2015). As a landlocked country, many Nepalese do not eat tuna, making it not essential; instead, they desperately needed rice and pulses.

Food/Drug Safety, Quality, and Appropriateness

Apart from issues associated with cultural and climatic appropriateness, food safety and quality is also a concern for food from foreign countries and even distant places within the disaster-affected country. Many authors in many countries have reported inappropriate food relief. Kent (1987) noted that dried skimmed milk and beans were distributed among Somalian refugees who were living in eastern Ethiopia in the late 1970s. The refugees traded these food items for more familiar foods or other items. Similarly, Henry (2000) reported that cornmeal and boxed breakfast cereal were distributed to West African refugees in the 1990s, whose staple is rice. Having no idea how to prepare cornmeal into recognizable "food," Sierra Leonean refugees were forced to sell the cornmeal to local markets in exchange for bags of rice. A similar situation was observed in Bangladesh after the country was struck by Cyclone Gorky in 1991. France sent frozen croissants and frozen dinners, and the USA sent a shipment of peanut butter for the cyclone victims. However, at the time, no Bangladeshis knew how to eat these items and ultimately gave these items to their animals as feed (Powell 2004).

Often foreign relief items are trapped in customs and do not get to disaster survivors when they can use them. There is also concern about the quality and safety of food items. For example, Cupples (2007) reported that after 1998 Hurricane Mitch in Nicaragua, relief agencies distributed beans among the survivors. However, these beans were so hard that they never cooked, despite using heaps of firewood. Following the 9/11 terrorist attacks in Manhattan, New York, surrounding areas donated food, but much of this food was not inspected to ensure safety and was not adequately refrigerated. Many responders refused to eat the food, concerned that ad hoc food donations could contain hazardous contaminants (Kendra and Wachtendorf 2004; Nelan 2016).

As in many other major disasters in the past, the United Nation's World Food Program (WFP) came forward to help meet the food needs of people affected by the earthquakes in Nepal. The WFP started to deliver emergency food supplies, including rice, pulses, and high-energy biscuits to earthquake survivors in Nepal beginning on April 29, 2015, using stocks that WFP already had in country from its existing operation before the earthquake. The WFP had a plan to provide food for 1.4 million people in urgent need of assistance over the next three months at a cost of $116.5 million (WFP 2015a). However, some survivors complained that the quality of rice provided by the WFP was poor, probably because the storage facilities were inadequate.[9] Some survivors did not use the rice while others fermented it to produce domestic alcohol, or mixed it with better quality rice and ate it, or gave it to their cattle (Paul et al. 2017).

The WFP authorities in Nepal initially denied distributing spoiled rice to earthquake survivors. The organization claims that it identified the problem before distribution began in Laprak, and bags containing spoiled rice were removed and replaced with good rice. Even so, the

WFP undertook an investigation of distribution of damaged rice among the earthquake survivors in June 2015. In late December 2015, the WFP disposed of a large quantity of damaged foodstuffs (including rice, peas, and biscuits) stored in the warehouse at Deurali in Gorkha district (*Kathmandu Post* 2015b).

In addition to rotten rice, the earthquake survivors of Kabhre district complained about the high percentage of broken rice grains in the rice. The WFP replaced the rice and discontinued using local suppliers of food items. In another case, the WFP received a shipment of yellow split peas in Nepalgunj that did not meet quality specifications. As soon as the problem was identified, the contract was canceled, and none of the food was distributed (*Kathmandu Post* 2015b).

In the immediate aftermath of natural disasters, moreover donated pharmaceuticals often pose serious threats to survivors. Often drugs past their expiration dates have been sent to disaster areas. For example, after the 1976 earthquakes in Guatemala, some drugs sent by foreign donors had expired in August 1934. Most drugs were not only useless; they arrived unsorted, unlabeled, poorly packaged, or were not intended for emergency use (Paul 2011). After the 1988 earthquake in northern Armenia, which killed some 25,000 people, foreign donors sent 5000 tons of drugs and medical supplies. Less than one-third of these supplies were immediately usable; 11% of the remaining were inappropriate for the situation, 8% had expired, and much of the remainder was inadequately labeled in as many as 21 different languages (Autier et al. 1990; Alexander 2000). Similarly, 37% of the medicine sent to El Salvador after the 2001 earthquakes was completely inappropriate for earthquake survivors (Holguin-Veras et al. 2014a).

The situation was no different in Bangladesh after a 1991 cyclone that killed 139,000 people. Rahman and Bennish (1993, 908) observed that much of the "… massive outpouring of drugs and medical supplies to the cyclone-affected areas … was not needed." Still worse, many essential drugs and supplies were not provided at all. After the 1970 cyclone in East Pakistan (now Bangladesh), newspapers around the world reported a major cholera outbreak in the affected areas (Kent 1987). In response, medicines poured into the country (including weight-reducing and contraceptive pills), and many cyclone survivors were vaccinated against cholera. However, Sommer and Mosley (1973) found no evidence of increased levels of cholera. This, by the way, does not necessarily show that the press is not concerned with the plight of people affected by disasters.

Too Many of the Same Items

Disaster survivors need some items urgently. But the authorities face serious problems if these items are not sent in sufficient quantities. After the 2010 Haiti Earthquake, water was in severely short supply. News media reported this shortage, which resulted in immense supply of bottled water, including energy drinks like Red Bull. Port-au-Prince was saturated with bottled water (Holguin-Veras et al. 2012b). As a result, however, shipments of other necessary supplies were halted for a number of days. Similarly, after the 2011 Fukushima disaster in Japan, too many blankets and items of clothing were shipped for the survivors (Nagurney 2017). Some of the donated items were useful, but these items were sent in large quantity and created storage problems. A considerable number of items sent from foreign countries were described in languages other than Japanese, which caused delays in handling by local authorities (Ito et al. 2014).

As opposed to convergence, a few studies (e.g., Ito et al. 2014; Paul et al. 2017) have reported a shortage of equipment at disaster sites. Equipment capable of handling the damage caused by the disaster was necessary, but no actors delivered equipment. Far too little equipment for handling debris arrived after the 2011 Japan Earthquake. Paul et al. (2017) reported a severe shortage of supply trucks and drivers to carry relief goods to remote areas of Nepal after the 2015 earthquake there. At the time of the first major earthquake in April 2015, Nepal had no earthmovers and cranes to clear roadblocks and start emergency relief operations, especially in remote areas. The U.N. had only three helicopters available to distribute aid, which is far too few for the size of the disaster and the need for relief. Helicopters are, moreover, also vulnerable to mountain terrain and weather conditions.[10]

The review of the relevant literature clearly suggests that among all major disasters occurring over the last several decades, survivors of the 2004 IOT received more redundant relief items from international actors than any other disaster, possibly because of huge geographic areas affected. All other extreme events affected part of a country and/or relatively small areas of several countries. Whatever the geographical extent of a disaster, international actors should be aware of both the human and physical geography of disaster-affected countries before sending emergency relief items for survivors.

Among major relief operations conducted in recent years, the number of inappropriate relief items distributed among the earthquake survivors in Nepal seems relatively low for two possible reasons. First, one month after the earthquake, the Government of Nepal imposed several rules and regulations for bringing relief items to the country, including customs inspections and import taxes (Weesjes 2015). Nepalese authorities argued that customs checks were necessary because they had no idea what was coming into the country. The international community criticized these actions, which delayed the disaster response, but the rules probably discouraged sending unnecessary items. For example, among other things, Singapore authorities collected old bras, expired medicine, and puzzle sets for Nepalese earthquake survivors but ultimately did not send them on to Nepal probably because of the import taxes (Han 2015).

Second, humanitarian aid organizations have had a strong presence in Nepal over the last few decades, largely responding to natural disasters like floods, landslides, and earthquakes, not to mention armed civil conflicts between the Government of Nepal (The Royal Nepal Army or RNA) and the Communist Party of Nepal (Maoists), conflicts that began in 1996 and ended in 2006. Many of these organizations have worked in collaboration with the Nepal government and Nepalese NGOs in the past to provide relief to the Nepalese people. These organizations even distributed relief resources to survivors of flooding and landslides occurring in 18 districts of Nepal in August 2014 (IASC 2015). Given this situation, it is surprising that these organizations did not better understand the immediate needs of earthquake survivors; still, the knowledge they did have must certainly have reduced the number of inappropriate number of items they may have received.

Note that these examples of material convergence represent a wide range of disasters that affected relatively large geographic areas as well as large populations. Such convergence problems also exist for disasters like tornadoes, which affect very small land areas. Simmons and Sutter (2011) reported that the mean damage area caused by all tornadoes from 1950 to 2007 differs by EF Scale category; EF-0 tornadoes damaged only 0.03 sq. mile on average. EF-5 category tornadoes have a damage area of only 9.86 sq. miles on average.[11] Holguin-Veras et al. (2014a) reported convergence of materials, however, after the 1953 Arkansas tornado and the 2011 Joplin, Missouri, tornado. They claimed that 60% of the donated items were not usable in the Arkansas disaster; about 70% of the clothing received by survivors of Joplin tornado was unusable. Only 10–15% of clothing donations were distributed among survivors.

In another example of a tornado, unnecessary items arrived from all parts of the USA as the national media broadcast the news of the EF-4 tornado that struck Harrisburg, Illinois, on February 29, 2012. Defying the often-used caution, "anything but clothes" or "A.B.C.," people sent huge quantity of clothes from across the country. Receiving and sorting these clothes took a massive amount of time and effort on the part of a limited number of disaster responders in this small community. Clothing was never critical to the tornado survivors. Rozdilsky (2012) reported a large reserve of used clothing in the state that could have filled any need in Harrisburg after the tornado. An overabundance of other items also arrived, such as large pallet of lock-sets. Material convergence was also reported after two major tornadoes (EF-4 and FE-5) occurred in May 2013 around Oklahoma City (Nelan 2016).

In its Fact Sheet 2013, USAID provided examples of five improper donation items sent to different countries between 1994 and 2010 (USAID 2013). Table 5.1 lists these items and countries.

Table 5.1 Examples of improper donations provided by USAID

Country/Region	Year	Disaster	Improper item
Rwanda	1994	Genocide	Weight loss drinks
Central America	1998	Hurricane Mitch	Chandeliers
Honduras	1998	Hurricane Mitch	Prom gowns
Afghanistan	2001	War	Pork (labeled as beef)
Haiti	2010	Earthquake	Fertility drugs

Source Prepared by the author based on USAID (2013, 3)

Conclusion and Recommendations

Among the three types of convergence, material convergence causes the most serious problems in disaster relief operations. The evidence suggests that international actors and donors still send inappropriate items as emergency aid, indicating a need for improved management of donations and the relief distribution system for future extreme events. Even transporting disaster relief items from distant places often costs more than the value of these items, and local capacities are often strained by the resources required to manage disaster relief activities. Moreover, sending redundant and inappropriate relief goods indicates a lack of geographic and cultural knowledge among those who donate items to survivors of disasters. Local actors and the governments of disaster-affected countries must discourage international actors from sending unnecessary relief items. Such items can be offensive to disaster survivors and may actually impede recovery efforts because unnecessary items disrupt the ability of responders to effectively manage and distribute items that are needed.

Authorities in disaster-affected countries bear some responsibility for not preparing a list of items survivors of extreme events need. Implementing procedures for rapidly assessing needs is useful in avoiding donations of unnecessary relief items. However, because such donations are likely the result of ignorance of the cultural and religious norms or of the climate of the affected country, organizing geographic and cultural awareness campaigns may be necessary for all entities providing emergency support, including international actors. The Red Cross and Red Crescent Society of individual countries can organize awareness campaigns and advise all actors about acceptable donations for disaster survivors in different cultures and climates.

Additionally, large relief organizations can organize focus group meetings for potential donors of NP supplies. Electronic and social media can also help international actors discover what items survivors of natural disasters need urgently, especially since social media has been used extensively in recent disasters to convey and receive information about friends, relatives, and family members.

In a paper on the response to the 2011 Tohoku Japan disasters, Holguín-Veras et al. (2014b) developed eight general recommendations for improving post-disaster response logistics. One set of recommendations focus on material convergence that could help minimize material convergence during future events:

1. Ensure that proper control procedures are in place to allow only high-priority supplies to enter the disaster area, divert low-priority supplies to secondary distribution centers (DCs) in the outskirts of the impacted areas, and refuse/destroy non-priority supplies;
2. Implement proactive donation management plans that provide potential donors and international actors with guidelines about what and when to donate;
3. Engage the media so that they convey a realistic representation of needs; and
4. Use information systems to try to match needs on the ground to potential donors.

Holguin-Veras et al. (2014a) also proposed using a schematic to identify entry points and end sites, which could help stop and/or reduce the flow of unnecessary relief items to disaster survivors. This schematic can be extended to non-affected countries as well. Figure 5.1 illustrates both disaster-affected and non-affected countries. For convenience, Fig. 5.1 shows only one entry/exit point in each affected and non-affected country. Entry point is in the affected country, while the

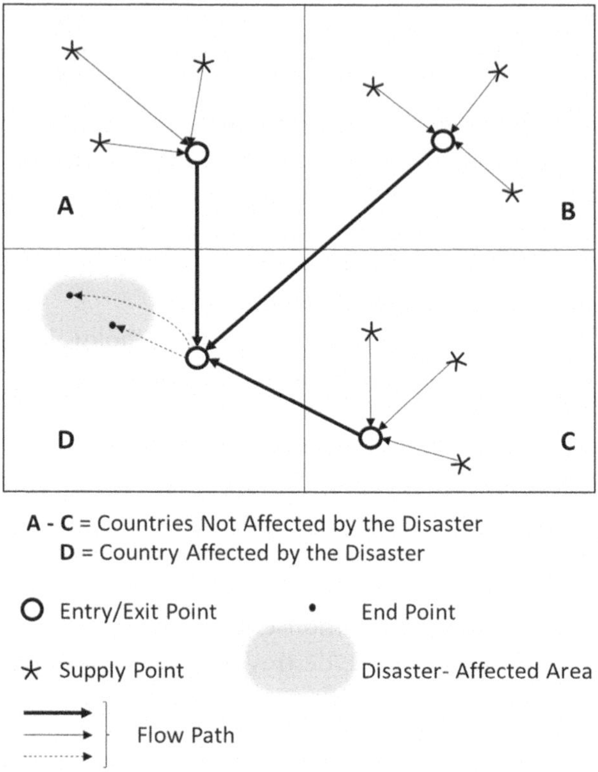

Fig. 5.1 Schematic of emergency supplies flowing to a disaster site (*Source* Modified after Holguin-Veras et al. (2014a))

exit point is in the non-affected countries. For the same reason, the figure shows only three sites in each country as supply points. These represent the people or all types of organizations who donate goods for disaster survivors. Holguin-Veras et al. (2014a) call these points donor sites. Once again, for convenience, Fig. 5.1 shows only two points as end points inside the disaster area.

An ideal entry/exit point would be the national capital of each country or alternatively, a major city. An entry/exit point allows a check for any unnecessary items being sent to disaster survivors of the affected country, and if such items are found, the appropriate authorities can discard the items immediately. The national Red Cross and Red Crescent Society or a relevant government agency could be responsible for this.

Two strategies could reduce dispersal of unnecessary relief items to survivors of future disasters. First, local actors and disaster-affected countries can quickly prepare a list of acceptable or necessary items for disaster survivors and send this list to international actors. International agencies, particularly large organizations, such as the American Red Cross and the Center for International Disaster Information, recommend sending cash donations, not in-kind donations, to avoid logistical challenges in providing disaster relief (American Red Cross 2015; Center for International Disaster Information 2015). For this reason, all disaster relief charities ask for cash donations. With cash, they can buy exactly what disaster survivors need, often from local sources. By doing this, money stays in the disaster-affected local economy and cuts down on the costs of shipping (Roos 2017). These agencies firmly believe that cash-based assistance can meet multiple needs across sectors, including food, non-food items, and services; it can also be used for recovery efforts.

Sending cash offers other potential advantages as well (Dreze and Sen 1989; Paul 2007, 2011). Cash donations do not occupy storage space, cost little to transport regardless of distance, and are valuable regardless of culture and society. Further, cash donations are rapidly and easily convertible and transferable (Cole-Corde 2013). In the 2005 Pakistan Earthquake response, cash grants successfully allowed affected households to meet their immediate needs (ALNAP 2008; Sanderson and Ramalingam 2015). However, sending cash does offer: fear of robbery, price hikes on essential items, and misuse of cash by disaster survivors (e.g., buying drinks instead of food). Some donors feel that cash donation is less personal and fails to establish any connection between donors and disaster survivors. By contrast, material donations are more likely to mean something to the survivors (Nelan 2016). Dudasilk (1982) claims that developed countries often deliberately donate cash as disaster relief because using their own materials, consultants, and contractors will cause up to as much as 70% of the cash to flow back to the country of origin.

The humanitarian community must work together to solve problems associated with convergence of unnecessary and useless materials to disaster sites. They must stress that those wishing to send relief materials to countries in need ought to consider the culture and climatic conditions of those countries. Because of cultural differences, an item may be unnecessary or even offensive to the people of one country, but may be acceptable to the people of another country. For example, sending beef was offensive to most Nepalese earthquake survivors but would be appropriate and acceptable to earthquake survivors of Muslim-dominated countries. This is an important point applicable to every disaster context. Proper planning and coordination will help emergency responders and managers avoid some of the negative consequences of convergence. Finally, the field requires more research to examine whether these recommended measures work regardless of the context in which they are applied.

Notes

1. Not all actors send disaster relief with good intentions; some relief is sent with ulterior motives. After the IOT, in the Colombo port, Sri Lankan officials seized several containers packed with narcotics, weapons, and other illegal goods sent as disaster relief by criminal elements from foreign countries (*Sikh News* 2005). Yet other donors, particularly private companies, use the disaster as a convenient mechanism to dump expired supplies or products that did not sell (Holguin-Veras et al. 2014a).
2. There is some debate about the reality of this myth. For example, Quarantelli (2007) reported a little evidence of looting behavior after disasters. In contrast, Frailing (2007, 1) claimed that "looting is not a myth, but a reality of disasters."
3. Government officials and NGO workers also determine the flow of disaster relief goods. For example, 2004 tsunami survivors in the Indian state of Tamil Nadu complained that officials and workers exaggerated reports of damage in some areas, while underestimating the devastation in others. As a consequence, many of those most in need of relief did not receive, while others who not even impacted by the tsunami enjoyed relief benefits (P'Rayan 2005).

4. The Haiti Earthquake was a turning point for how social media is used in disaster situations. Until 2010, social media was independently evolving, but the size and inherent emotional appeal of that disaster created the right environment for it to flourish.
5. Amanda et al. (2008) call social media online social convergence.
6. Ushahidi is a nonprofit software company that develops free and open-source software (LGPL) for collecting, visualizing, and interacting with information, using mapping on mobile phones. Ushahidi, which translates as "testimony" from Swahili, was developed in 2008 in Kenya after post-election violence, but was used for mapping disaster sites in the immediate aftermath of the 2010 Haiti Earthquake. Other cloud mapping sources used in Haiti are: CrisisCamp Haiti and OpenStreetMap (OSM) (Zook et al. 2010).
7. This information was collected by the author through focus group meetings held in Nepal in January 2016 (Paul et al. 2017). These meetings were supported by the National Science Foundation's RAPID (Award Number 1547554) Grant to study how linking networks contributed to disbursing emergency assistance among survivors of the 2015 Nepal Earthquakes.
8. Arnette and Zobel (2016) reported that messaging focused on encouraging financial donations and proper types of material donations, which together with effective donations management, reduced material convergence. A key aspect of messaging was the development of the HelpColoradoNow website, a project of Colorado Voluntary Organizations Active in Disasters (COVOAD) in partnership with the Donation and Volunteer Coordination Team (DVCT) and the Colorado Department of Homeland Security and Emergency Management (DHSEM). Not only was Aidmatrix used on HelpColoradoNow to control the list of acceptable items for donation, but Aidmatrix could screen offers, which was very important to reducing the spontaneous flow of donations arriving at the distribution site within the flood-affected areas (Arnette and Zobel 2016).
9. The distribution of WFP's broken or spoiled rice had never been reported before the incident in Nepal. The WFP identified three reasons this may have happened: The rice could have been received damaged from the supplier; the rice could have been damaged because of high temperatures in warehouses; or the rice could have been damaged during transportation (WFP 2015b). According to the WFP, in Nepal,

the quality of food for emergency distribution depends on the WFP, suppliers, transporters, partners like the Nepal Red Cross Society, independent inspection services and laboratories, and the Government of Nepal. All suppliers of WFP food, both local and international, must comply with international food safety and quality requirements, and their performance is evaluated regularly. Once food is purchased its quality and safety are assessed against international standards.

10. On May 12, 2015, a US military helicopter crashed, killing six US Marines. On June 2, 2015, a commercial helicopter crashed took lives of four aid workers (*Kathmandu Post* 2015c).
11. Intensity of tornado was measured on the Fujita scale (F-Scale) from 1971 through 2007. This scale was introduced by Theodore Fujita in 1971 and it ranged from F-0 (light damage) through F-5 (incredible damage). This scale was revised in 2007 and since then it is known as the Enhanced Fujita scale or EF Scale. The EF Scale has the same basic design as the F-Scale (Paul 2011).

References

Abbasi, M., and M.H. Salehnia. 2013. Disaster Medical Assistance Teams After Earthquake in Iran: Propose a Localized Mode. *Iranian Medical Red Crescent Medical Journal* 15 (9): 829–835.

ABE Streep. 2015. Nepal's Aid System is Broken. So These Lifesavers Hacked It. http://www.wired.com/2015/05/nepal-earthquake-aid. Last Accessed 24 Feb 2016.

ALNAP (Active Learning for Accountability and Performance in Humanitarian Action). 2008. *Responding to Earthquakes 2008: Learning from Earthquake Relief and Recovery Operation.* London.

Alexander, D. 2000. *Confronting Catastrophe.* Oxford: Oxford University Press.

Alexander, D.E. 2014. Social Media in Disaster Risk Reduction and Crisis Management. *Science and Engineering Ethics* 20 (3): 717–733.

Amanda, L.H., L. Palen, J. Sutton, S.R. Liu, and S. Vieweg. 2008. "Site-Seeing" in Disaster: An Examination of On-Line Social Convergence. In *Proceedings of the 5th International ISCRAM Conference*, May, Washington, DC.

American Red Cross. 2015. Donating Goods. http://www.redcross.org/support/donating-fundraising/other-ways-to-give/donating-goods. Last Accessed 8 July 2017.

Arnette, A.N., and C.W. Zobel. 2016. Investigation of Material Convergence in the September 2013 Colorado Floods. *Natural Hazards Review* 17 (2): 05016001.

Autier, P., M.C. Ferir, A. Hairapetien, A. Alexanian, V. Agoudiian, G. Schmets, G. Dallemagnew, M.N. Leva, and J. Pinel. 1990. Drug Supply in the Aftermath of the 1988 Armenian Earthquake. *Lancet* 335: 1388–1390.

Bemon, B.M., and B. Blacik. 2008. Performance Measurement in Humanitarian Relief Chains. *International Journal of Public Sector Management* 21 (1): 4–25.

Benglsson, L., X. Lu, A. Thorson, R. Garfield, and J. von Schreeb. 2011. Improved Response to Disasters and Outbreaks by Tracking Population Movements with Mobile Phone Network Data: A Post-earthquake Geospatial Study in Haiti. *PLoS Medicine* 8 (8): 1–8.

Bradshaw, S. 2013. *Gender, Development and Disasters*. Northampton, MA: Edward Elgar.

Brenhouse, H. 2011. China's Great Salt Rush: Nuke Fears Cause Supermarket Swarms. *Time*. http://newsfeed.time.com/2011/03/17the-great-salt-rush-nuke-fears-have-chinese-swarming-supermakets. Accessed 23 Sept 2016.

Brown, G. 2005. Making a Tragedy Out of a Disaster. *The World Today* 61 (1): 7–8.

Center for International Disaster Information. 2015. *Guidelines for Giving*. USAID. http://www.cidi.org/how-disaster-relief-works/guidelines-for-giving/#VP2t2mQ1fLU. Last Accessed 8 July 2017.

Chia, E.S. 2007. Engineering Disaster Relief. *IEEE Technology and Society Magazine*, 24–29.

Cole-Corde, A.N. 2013. Informal Material Convergence Behavior and Emergency Management: Looking a Gift Horse in the Mouth. Master's thesis, North Dakota State University, Fargo, ND.

Cupples, J. 2007. Gender and Hurricane Mitch: Reconstructing Subjectivities after Disaster. *Disasters* 31 (2): 155–175.

Daniel, F.J., and K. Mahr. 2015. Nepal, Aid Agencies Trade Blame as Confusion Mars Quake Relief. http://in.reuters.com/article/earthquake-nepal-aid-idINKBNONQ1DG20150505. Last Accessed 2 Apr 2016.

Dewaal, A. 2017. On the Significance of the Declaration of Famine in South Sudan. World Peace Foundation. http://sites.tufts.edu/reinventing-peace/2017/02/28/on-the-significance-of-the-declaration-of-famine-in-south-sudan/. Last Accessed 1 June.

Dreze, J., and A. Sen. 1989. *Hunger and Public Action*. Oxford: Clarendon.

Dudasilk, S. 1982. Unanticipated Repercussions of International Disaster Relief. *Disasters* 6: 31–37.

Elia, G. 1928. *The International Assistance Union.* Cambridge: Geographical Congress of Cambridge.

Express Tribune. 2015. No Beef Content in Food Dispatched by Pakistan to Nepal. Apr 30.

Frailing, K. 2007. The Myth of a Disaster Myth: Potential Looting Should be Part of Disaster Plans. *Natural Hazards Observer* 31 (4): 3–4.

France, D., and F. Ordonez. 2002. $75 Million of Staff. *Newsweek* 139 (7): 62–63.

Fritz, C.E., and J.H. Mathewson. 1957. *Convergence Behavior: A Disaster Control Problem.* Washington, DC: National Academy of Science.

Gautham, S. 2006. Teach the Girls to Swim Tsunami: Survival and the Gender Dimension. http://www.countercurrents.org.gen-gautham100806htm. 2 Apr 2007.

Han, L.Y. 2015. Some Singaporeans Donated Items 'Of No Use' to Nepal Earthquake. *The Straits Times*, May 6.

Henry, D. 2000. Embodied Violence: War and Relief Along the Sierra Leone-Guinea Border. Unpublished PhD thesis presented to Southern Methodist University, Dallas.

Holguin-Veras, J. 2012. The Donations Sandy's Victim's, Victims Don't Need Spanish Flags, Tuxedos, Sex Toys: Inappropriate Donations Inevitably Pour in After Disasters. *Los Angeles Times*, Nov 3.

Holguin-Veras, J., N. Perez, S. Ukkusuri, T. Wachtendorf, and B. Brown. 2007. Emergency Logistics Issues Impacting the Response to Hurricane Katrina: A Synthesis and Preliminary Suggestions for Improvement. *Transportation Research Board* 2022 (2022): 76–82.

Holguin-Veras, J., M. Jaller, L.N.V. Wassenhove, N. Perez, and T. Wachtendorf. 2012a. On the Unique Features of Post-disaster Humanitarian Logistics. *Journal of Operational Management* 30 (7–8): 494–506.

Holguin-Veras, J., M. Jaller, and T. Wachtendorf. 2012b. Comparative Performance of Alternative Humanitarian Logistic Structures After the Port-au-Prince Earthquake: ACEs, PIEs, and CANs. *Transportation Research Part A* 46: 1623–1640.

Holguin-Veras, J., M. Jaller, N. Luk, V. Wassenhove, N. Perez, and T. Wachtendorf. 2014a. Material Convergence: Important and Understudied Disaster Phenomenon. *Natural Hazards Review* 15 (1): 1–12.

Holguin-Veras, J., E. Taniguchi, M. Jaller, F. Aros-Vera, F. Ferreira, and R.G. Thompson. 2014b. The Tohoku Disasters: Chief Lessons Concerning the Post-disaster Humanitarian Logistics Response and Policy Implications. *Transport Research Part A* 69: 86–104.

IASC (Inter-Agency Standing Committee) Reference Group for Mental Health and Psychosocial Support in Emergency Settings. 2015. *Nepal Earthquake 2015: Desk Review of Existing Information with Relevance to Mental Health & Psychosocial Support*. Kathmandu.

IFRC (International Federation of Red Cross and Red Crescent Society). 2005. *World Disaster Report*. Geneva.

IFRC (International Federation of Red Cross and Red Crescent Society). 2006. *World Disaster Report 2006: Focus on Neglected Crises*. Geneva.

IFRC (International Federation of Red Cross and Red Crescent Society). 2013. *World Disasters Report: Focus on Technology and the Future of Humanitarian Action*. Geneva.

Ito, H., W. Wisetjindawat, and M. Yokomatsu. 2014. Improving the Operational Efficiency of Humanitarian Logistics in the Aftermath of a Large-Scale Disaster. *Journal of Integrated Disaster Risk Management* 4 (2): 142–155.

Kathmandu Post. 2015a. Government to Conduct Health Camps in Quick-Hit Districts. Dec 30.

Kathmandu Post. 2015b. WFP Destroys Rotten Food Items. Dec 30.

Kathmandu Post. 2015c. Four Dead in Sindhupalchock Chopper Crash. June 2.

Kendra, J.M., and T. Wachtendorf. 2004. *Rebel Food …Renegade Supplies: Convergence After the World Trade Center Attack*. Newark, DE: Disaster Research Center, University of Delaware.

Kent, R.C. 1987. *The Anatomy of Disaster Relief: The International Network in Action*. London: Frances Printer.

Letukas, L. 2008. Is Aid A Social Problem? Cross-National Media Construction of Relief Efforts Following the Indian Ocean Tsunami. Master's thesis, University of Delaware, Newark.

Lindsay, B.R. 2011. *Social Media and Disasters: Current Uses, Future Options, and Policy Consideration*. Washington, DC: Congressional Research Service Report for Congress.

Lu, X., D.J. Wrathall, P.R. Sundsey, M. Nadiruzzaman, E. Wetter, A. Iqbal, T. Qureshi, A. Talem, G. Canright, K.E. Monsen, and L. Benglsson. 2016. Unveiling Hidden Migration and Mobility Patterns in Climate Stressed

Regions: A Longitudinal Study of Six Million Anonymous Mobile Phone Users in Bangladesh. *Global Environmental Change* 38: 1–7.

Mahmud, T., and M. Prowse. 2012. *Corruption in Cyclone Preparedness and Relief Efforts in Coastal Bangladesh: Lessons for Climate Adaptation?* Antwerpen, Belgium: University of Antwerp.

Mail Today. 2015. Pakistan Severs Beef to Nepal Earthquake Survivors. Apr 30.

McEntire, D.A. 1999. Issues in Disaster Relief: Progress, Perpetual Problems and Prospective Solutions. *Disaster Prevention and Management: An International Journal* 8 (5): 351–361.

McKay, B.J., n.d. Managing Volunteer Organizations in Time of Disaster. Unpublished Paper, Adelphi University, New York.

Merchant, R.M., S. Elmer, and N. Lune. 2011. Integrating Social Media into Emergency-Preparedness Efforts. *New England Journal of Medicine* 385: 289–291.

Miller, G. 2005. The Tsunami's Psychological Aftermath. *Science* 209: 5737.

Nagurney, A. 2017. How Disaster Relief Efforts Could Be Improved with Game Theory. *The Conversation*, Mar 9. http://theconversation.com/how-disaster-relief-efforts-could-be-improved-with-game-theory-72923. Last Accessed 30 June 2017.

Natural Hazards Observer. 2011. Looking at the Second Life of Social Media. 33 (2): 2–3.

Neal, D. 1994. The Consequences of Excessive Unrequested Donations: The Case of Hurricane Andrew. *Disaster Management* 6 (1): 23–28.

Nelan, M.M. 2016. The Social Construction of Donations: Agility, Adaptability, and Alignment as Success Determinants in Relief Supply Chains. A PhD dissertation, University of Delawar, Newark.

PAHO (Pan American Health Organization). 2001. *Establishing a Mass Casualty Management System,* 2nd ed. Washington, DC: PAHO.

PAHO. 2003. *WHO-PAHO Guidelines for the Use of Foreign Field Hospitals in the Aftermath of Sudden-Impact Disaster*. San-Salvador, El-Salvador: WHO.

Paul, B.K. 2006. Disaster Relief Efforts: An Update. *Progress in Development Studies* 6 (3): 211–223.

Paul, B.K. 2007. 2004 Tsunami Relief Efforts: An Overview. *Asian Profile* 35 (5): 467–478.

Paul, B.K. 2011. *Environmental Hazards and Disasters: Contexts, Perspectives and Management*. Hoboken, NJ: Wiley-Blackwell.

Paul, B.K., B. Acharya, and K. Ghimire. 2017. Effectiveness of Earthquakes Relief Efforts in Nepal: Opinions of the Survivors. *Natural Hazards* 85: 1169–1188.

Plan International. 2005. *Children and the Tsunami*. Bangkok: Plan Ltd.

Peng, C. 2005. Tsunami Survivor Receiving Lopsided Relief, Report Says. http://www.dailycal.org/article.php?id.18842. 23 Mar 2011.

Powell, B. 2004. Where Should Your Money Go? *Time* 165 (2): 44–45.

P'Rayan, A. 2005. Impact of Tsunami Relief and Rehabilitation Work in India. http://www.Worldpress.org/Asia2116.cfm. Last Accessed 21 June 2010.

Prince, S.H. 1920. Catastrophe and Social Change, Based Upon a Sociological Study of the Halifax Disaster. PhD dissertation, Columbia University, New York.

Quarantelli, E.L. 2007. The Myth and the Realities: Keeping the "Looting" Myth in Perspective. *Natural Hazards Observer* 31 (4): 2–3.

Rahman, M.O., and M. Bennish. 1993. Health Related Response to Natural Disasters: The Case of the Bangladesh Cyclone of 1991. *Social Science and Medicine* 36: 903–914.

Roos, D. 2017. 5 Donation Tips for (Really) Helping Disaster Victims, Posted on September 18. https://money.howstuffworks.com/5-donation-tips-for-really-helping-disaster-victims.htm. Last Accessed 17 Nov 2017.

Rozdilsky, J.L. 2012. *Disaster Management with Limited Local Resources: The 2012 Illinois Leap Day EF-4 Tornado*. Quick Response #244. Boulder, CO: The Natural Hazards Center, University of Colorado, Boulder.

Sanderson, D., and B. Ramalingam. 2015. *Nepal Earthquake Response: Lessons for Operational Agencies*. ALNAP Lessons Paper. London: ALNAP/ODI.

Saxena, A. 2015. Nepal Earthquake: Pakistan Ducks After Beef Relief Blunder. *India Today*, May 1.

Sikh News. 2005. Sikh Donations Left Rot at Port in Sri Lanka. Aug 4. http://www.sikhsangat.org/publishg/article_237.shtm. Last Accessed 8 Aug 2005.

Simmons, K.M., and D. Sutter. 2011. *Economic and Societal Impacts of Tornadoes*. Boston: American Meteorological Society.

Sommer, A., and W. Mosley. 1973. The Cyclone: Medical Assessment and Determination of Relief and Rehabilitation Needs. In *Disaster in Bangladesh: Health Crises in a Developing Nation*, ed. L. Chen. New York: Oxford University Press.

Spencer, R. 2010. Islamic Supremacists Serve Hindus Beef in Pakistan Flood Camps. *Jihad Watch*, Aug 24.

Telegraph. 2015. UN: Customs Rules Delaying Foreign Aid Nepal: Send Grains, Not Tuna & Mayo. May 2.

Telford, J., and J. Cosgrave. 2007. The International Humanitarian System and the 2004 Indian Ocean Earthquake and Tsunamis. *Disasters* 3 (1): 1–28.

Tierney, K. 2003. *Conceptualizing and Measuring Organizational and Community Resilience*. Newark: University of Delaware Disaster Research Center.

Trainor, J., and J. Kendra. 2005. More Than a Wave: Exploring the Social Impacts of the Indian Ocean Tsunami. *Natural Hazards Observer* 29 (5): 1–3.

Troutman, E. 2015. What Happened to the Aid? Nepal Earthquake Response Echoes Haiti. *Aid Works*, June 19. http://aid.works/2015/06/nepal-haiti/. Last Accessed 8 June 2017.

UN (United Nations). 2005. Regional Workshop on Lessons Learned and Best Practices in the Response to the Indian Ocean Tsunami: Report and Summary of Main Conclusion. Medan, Indonesia.

UN–ESC (United Nations–Economic and Social Council). 2005. *Strengthening Emergency Relief: Rehabilitation, Reconstruction, Recovery and Prevention in the Aftermath of the Indian Ocean Tsunami*. New York. http://pdf.usaid.gov/pdf_docs/pdacy077.pdf.

USAID (US Agency for International Development). 2013. USAID Donations Outreach. http://pdf.usaid.gov/pdf_docs/pdacy077.pdf. Last Accessed 17 Nov 2017.

Wachtendorf, T. 2008. The Social Impact of May, 2008 Wenchuan China Earthquake. Report submitted to the Earthquake Engineering Research Institute.

Wachtendorf, T., J.M. Kendra, H. Rodriguez, and J. Trainor. 2006. The Social Impacts and Consequences of the December, 2004 Indian Ocean Tsunami: Observation from India and Sri Lanka. *Earthquake Spectra* 22 (3): 693–714.

Wachtendorf, T., B. Brown, J. Hologuin-Veras, and S. Ukkusuri. 2010. *Catastrophe Characteristics and Their Impact on Critical Supply Chains: Problematizing Material Convergence and Management Following Hurricane Katrina*. Newark: Disaster Research Center, University of Delaware.

Wachtendorf, T., S. Penta, and M.M. Nelan. 2015. When Push Comes to Shove: The Framing of Need in Disaster Relief Efforts. In *Learning and Calamities: Practices, Interpretations, Pattern*, ed. H. Enger, M. Schorch, and M. Voss, 235–252. New York: Routledge.

Weesjes, E. 2015. From the Editor. *Natural Hazards Observer* 39 (5): 3.

WFP (World Food Program). 2015a. WFP Rushes Food and Relief Supplies to Nepal Earthquake Survivors. *Kathmandu*, Apr 28.

WFP. 2015b. WFP Launches Investigation into Damaged Rice in Nepal. *Kathmandu*, June 19.

Yonder, A. 2005. *Women's Participation in Disaster Relief and Recovery*. New York: Population Council.

Zakaria, R. 2015. The Deceptions of Disaster Relief in Nepal: The Aid-Industrial Complex is in Dire Need of Overhaul. *Aljazeera Americaq*, July 2. http://america.aljazeera.com/opinions/2015/7/the-deceptions-of-disaster-relief-in-nepal.html. Last Accessed 8 June 2017.

Zook, M., M. Graham, T. Shelton, and S. Gorman. 2010. Volunteered Geographic Information and Croudsourcing Disaster Relief: A Case Study of the Haitian Earthquake. *World Medical & Health Policy* 2 (2): 7–33.

6
Conclusion

Disaster relief should represent a response that is immediate to alleviate suffering and save the lives of those affected by an extreme natural and/or man-made event. Ideally, it consists of fast-paced, apolitical, reactive crisis interventions of short-term duration to meet immediate subsistence needs of disaster victims and augment their disaster recovery process. Although disaster relief is not aimed to reduce the probability of future disasters, it does help survivors smooth the transition from necessary and immediate relief to long-term rehabilitation and redevelopment, and to return as quickly as possible to pre-disaster life. However, providing emergency aid to disaster survivors is full of complexities, largely a result of many changes in disaster relief operations over the past several decades. Although these complexities have made relief efforts very difficult and challenging, this trend also reflects that the field of disaster relief has grown tremendously over time. Not only have new humanitarian actors emerged at all levels, but also new terms and concepts have surfaced, which have matured the field.

In contrast to this development, disaster relief efforts seem to be confronted by many perpetual problems and challenges. It appears that experience and lessons learned from past emergency relief operations

B. K. Paul, *Disaster Relief Aid*,
https://doi.org/10.1007/978-3-319-77282-0_6

either have not resulted in improved performance in more recent disaster events, or such learning has been short-lived. In reviewing the nature of international emergency relief operations undertaken over past decades, Randolph Kent (1987) identified three interrelated characteristics: appropriate intervention, timely intervention, and coordination by relief entities. By appropriate intervention, he meant appropriate supplies being sent to disaster victims, supplies suited to geographic and cultural norms of the disaster-affected areas, thereby avoiding the dumping of useless and inappropriate relief items that has become a norm, as Kent noted. About timely intervention, Kent wrote, "if aid is not received in a timely manner, then it is no longer appropriate. Timely intervention involves not only an awareness of when aid is needed but also when aid is no longer needed" (Kent 1987, 25). Finally, coordination among people and organizations involved in relief operations is crucial because uncoordinated relief efforts result in duplication of aid, double deliveries to one community and none to another, and competition among providers of disaster relief. In addition to these factors, effectiveness of disaster aid distribution also depends on adequacy and fairness of emergency supplies to those who are severely affected by an extreme event (also see Chapter 3 of this book).

Many manuals, guides, evaluations and assessment reports, and academic analyses have already been published over the years to show how to avoid arbitrary relief assistance and to permit targeted relief to the most vulnerable populations. Moreover, the U.N. and international communities have also made vigorous attempts to make each disaster relief operation successful and effective. Unfortunately, these attempts have not achieved much success. This final chapter outlines the major problems and challenges of international disaster relief operation and forwards recommendations about how to improve future disaster relief efforts.

Before focusing on particular problems and challenges, this chapter addresses events that occurred in disaster relief operations over the last several decades. Knowing these events is essential to understand the complexity and dynamics of the relief process, especially as many of these events have had a positive impact on the process. Notable among them are the formation of the Office of Coordination of Humanitarian

Affairs (OCHA) by the United Nations in 1991 (also called UNOCHA or UN OCHA), the Inter-Agency Standing Committee (IASC) in 1992, the Active Learning Network for Accountability and Performance (ALNAP) in 1997, the Financial Tracking System (FTS) in 2000, the Good Humanitarian Donorship (GHD) in 2003, the Consolidated Appeal Process (CAP) in 2006, and the Global Humanitarian Platform (GHP) in 2007.[1] Additionally, the Cluster Approach in 2005 and Post-Disaster Needs Assessment (PDNA) in 2008, followed by the rise of social media/networking in the mid-2000s, all represent positive changes to improve global disaster aid operations.

The OCHA was established to coordinate the U.N. human response to natural disasters and complex emergencies. Before its formation, two U.N. organizations (the Department of Humanitarian Affairs, DHA, and its predecessor, the Office of the United Nations Disaster Relief Coordinator, UNDRC) were responsible for performing similar tasks. In 1998, due to reorganization, DHA merged into OCHA and was designed to be the U.N. focal point on major disasters (http://www.unocha.org/about-us/who-we-are). The OCHA's activities also include organizing and monitoring humanitarian funding, as well as generating information exchange, coordination, and rapid-response teams for emergency relief.

The ALNAP in humanitarian action is a London-based unique sector-wide network in the international humanitarian system comprising key humanitarian organizations and leading experts in the field. The broad range of experience and expertise from across the membership is at the heart of ALNAP's efforts to improve humanitarian performance through learning and accountability. ALNAP Full Members include agencies and individuals from five key constituencies that make up the international humanitarian sector: (1) bilateral and multilateral donors, (2) U.N. agencies, (3) the Red Cross and Red Crescent Movement, (4) international and national NGOs and NGO networks/umbrella bodies, and (5) academic establishments, research institutions, and independent consultants. The ALNAP members work together to identify common approaches to improved performance, to explore new ways to improve learning and accountability, and to share ideas, lessons, and innovations (Taylor et al. 2012).

The Financial Tracking Service (FTS) is a global, real-time database that records all reported international humanitarian aid contributions (including NGOs and the Red Cross/Red Crescent Movement, bilateral aid, in-kind aid, and private donations). FTS focuses on humanitarian response plans (HRPs) and refugee response plans (RRPs) because service covers the major humanitarian crises and because their funding requirements are well defined. FTS is managed by the UNOCHA, and all its data are provided by donors or recipient organizations.

Specifically, two positive events that occurred in disaster relief distribution deserve special mention. For the first time in the history of disaster relief operation, women and children received special attention from relief agencies after Mozambique experienced a devastating flood in 2000 and Bangladesh was hit by Cyclone Sidr in 2007, respectively. For the Mozambique flood, the USA distributed the resettlement cash grant to each affected family (USAID 2002). However, the grant was delivered to the woman of each flood-affected household. The woman was deliberately selected with the intention that she would utilize the money in the best interests of the whole household. This effort was a great success in coping with the flood impacts.

Relief agencies repeated the approach after the Haiti earthquake in 2010 by opening a disaster relief distribution center for women only. Although this initiative increased violence against women, it was undoubtedly a noble gesture. While some women were robbed or raped while returning to their homes with relief goods (Hebblethwaite 2011), with proper protection, this provision should continue in all future relief efforts.

Immediately following the Cyclone Sidr in Bangladesh, UNICEF together with Save the Children established 220 shelters for child survivors and distributed food and safe drinking water to children. They also provided psychosocial support and organized recreation for school-going children (Islam and Walkerden 2015). In fact, special distribution centers for the elderly should also be opened.

Another positive event associated with emergency response and relief operations is that many disaster survivors now have the ability to communicate their needs to both domestic and international responders through cellular connectivity and social media. In fact, since the

2010 Haiti earthquake, a new stakeholder has emerged and members of this stakeholder are called volunteered and technical communities (V&TCs). Members of V&TCs have been playing a major role in disaster relief efforts by providing free maps, remote sensing images through OpenStreetMap, Crisis Mapping, social media, and mobile devices (IFRC 2013).

Problems and Challenges

Note that the materials included in this and subsequent two sections are related to each other; thus, recommendations and future research suggestions come directly from problems and challenges. To begin, central to disaster relief operations is the distribution of emergency supplies among the most afflicted. Review of the relevant examples in previous chapters clearly suggests that not much progress has been achieved over the last several decades in disaster relief efforts, particularly equitable, uneven, or unbiased relief distribution in disaster-affected communities at the household level. This is so despite impartiality and neutrality being two of the three stated principles of the United Nations for all humanitarian actions. Violations of these principles should not be acceptable to anyone because such violations result in human rights abuse and introduce bribery and corruption into disaster relief disbursement.

Unfortunately, no concerted initiative has occurred in the past either to improve disaster aid distribution among the survivors of extreme events or to reach out to the most vulnerable groups, or to those who had suffered the greatest damage. One way to reach such groups is by private and public relief organizations targeting specific types of members as in the cases of children in Cyclone Sidr in Bangladesh and of women after the Haiti earthquake. Alternatively, the appropriate government authorities quickly identify and classify beneficiaries affected by a particular disaster for emergency response. Most vulnerable households, including those in remote or isolated locations, ethnic minorities, and socially disadvantaged (e.g., families without any income, families headed by women, families with person(s) with disability) should receive the highest priority for disaster relief. However, currently, little is done to collect data of vulnerable and non-vulnerable groups, yet specific disaster relief data about such groups could provide a rich narrative regarding the success of relief efforts for such survivors.

Next, highly dynamic and uncertain disaster environments present many overlapping challenges to timely distribution of emergency relief supplies to victims: rigid bureaucracy; absence of adequate transportation infrastructure and damage to available infrastructure; poor agency coordination/collaboration; lack of training and equipment to clear rubble, road blockages and to unload relief goods at entry points; shortage of fuel, transport vehicles, drivers, and labor; and few distribution points. Irrespective of level of development of disaster-affected countries, these seem to be problems common to almost all disaster relief operations for timely delivery of emergency supplies to the people who most urgently need them.

Moreover, lack of local knowledge and awareness, tax incentives, prosocial behavior of donors, disposal of unwanted and unnecessary household goods, and pleasing donors are some of the principal reasons for excess, unnecessary, and even useless material convergence at disaster sites. These donations clog up the supply chain, taking up space that could be used for essential items. Additionally, the expense of transporting these goods most frequently is not cost-effective, and invariably convergence of responders, information, and materials delays distribution of disaster relief in a timely manner and instead increases the suffering of survivors.

Another serious problem associated with the relief environment, particularly in developing countries, is that it is unregulated, which implies that no single organization is responsible to lead and coordinate emergency aid operations. Typically, the governments of the affected countries are responsible for such operations, and other actors are obliged to abide by their leadership (Balcik et al. 2010). With the emergence of many bilateral and multilateral relief organizations and agencies over time, such governments have been losing their authority to conduct relief operations. Another reason for this is that these governments do not have the necessary experience and knowledge to manage emergencies effectively, nor are they always accountable or transparent. These issues are making governments either non-functional or dysfunctional in their own disaster relief environments.

Recommendations

Local communities are the first responders to emergencies, and they are key actors in disaster risk management and mitigation. Therefore, their capacity for disaster relief should enlarge in the foreseeable future. Also, a participatory approach to relief distribution needs to be adopted by defusing local power relations and ensuring inclusion of marginalized groups. Such an approach contrasts with the more traditional top-down approach where needs of the disaster survivors are frequently assumed and "filled in" by officials, workers, and individuals involved in emergency relief operations. On the contrary, active participation in and opinions about all post-disaster activities by survivors will help guarantee the smooth and fair distribution of future disaster relief.

Three decades ago, Kent (1987, 14) wrote, "Relief to be effective, must be sensitive to local conditions and values." Therefore, adequate knowledge about local communities is essential to effectively and efficiently distribute emergency relief aid to disaster victims. This requires understanding and sometimes deferring to the larger social and organizational cultures that may hinder sustainable development (Henry 2005). In contrast to the traditional command and control approach to disaster relief efforts, actors involved in such efforts must seek local inputs, use local skills and technologies, and respect local cultural norms. This will further help to strengthen local disaster response capacities and thus increase disaster resiliency (Henry 2005).

Local occupational groups (e.g., farming and fishing) have the opportunity to provide important emergency support for those who are affected the most by extreme natural and man-made events.[2] Islam and Walkerden (2015) maintain that these occupational groups could contribute to disaster relief in several ways. First, they can function as self-help groups and provide local volunteers that will aid in the disaster response. Second, because they are derived from the local area, members of these organizations have accurate information about the people who are most affected and accordingly they can help in proper disbursement of disaster relief materials. Finally, these groups can also help the multitude of relief agencies identify the most vulnerable households and assist in identifying and contacting other groups (e.g., cooperatives) or individuals (e.g., political leaders and local philanthropists) who could help with disaster response.

Several researchers like Flint and Goyder (2006) observed that in response to the Flash Appeal, such early funding is not always based on need of disaster survivors primarily because typically incomplete damage assessment at this stage is often slow, imprecise, overlapping, and poorly shared. Rather, such assessment is often based on media reports and/or perceptions of government officials, which do not reflect reality. Besides, pledged amounts are often influenced by media-driven completion and political considerations as was the case between Japan and China in the context of the 2004 IOT. A Flash Appeal, therefore, should be based on an objective assessment of loss and damage caused by a particular disaster, and the pledged amount should not be based on perceived need of the affected people. Also, donors do not always know where their donations are going, or what population they serve, but such information should be available to donors.

Further, emphasis should be placed on preventing unnecessary and non-priority items that hinder relief flow. For those donating and collecting donations, collecting more items than are needed comes at little cost (aside from transportation to the affected area) and the reward of having donated altruistically is highly valued. However, if pleasing the donors is the main goal, the problems associated with convergence of massive amounts of unwanted materials will never be solved. Donors must be educated about the items that are high priority and necessary versus those that will clog the relief flow.

In Chapter 5, recommendations suggest how to reduce the flow of low-priority and unnecessary relief items to disaster sites. For donations coming from foreign countries, one way to stop such flow is to not allow low-priority relief supplies at the exit points (i.e., airports or seaports). Such points should enact the proper control procedures so that only high-priority supplies can leave the country. Similarly, at entry points, customs authorities and government officials of the disaster-affected country can check incoming relief goods. Although this will slow the flow of disaster relief items to survivors, this policy was successful in stopping and/or reducing flow of non-priority items during the relief phase after Nepal was hit by the 2015 earthquakes. Nepali authorities also imposed import taxes on many incoming relief items from foreign countries, which discouraged foreigners from sending unnecessary relief items to Nepal (Paul et al. 2017).

In addition, disaster-affected countries need to prepare a list of emergency items that disaster victims need either prior to disasters or immediately after and distribute it to donors and other international actors. The list should be disaster specific, particularly if it is prepared before the disaster. For example, tents are in great demand for survivors of earthquakes, but not so for flood or tornado victims. Similarly, farmers are generally affected by inland flooding in developing countries, while fishermen are affected by coastal flooding or tropical cyclones. To put a finer point on this, farmers will benefit most if they are provided with lost agricultural equipment, while fishermen will benefit if they receive help to build their damaged boats. In contrast, some relief items (e.g., water, medicine, and dry food) are needed by victims of all types of disasters. Also, relief agencies such as Oxfam and the Federation of Red Cross and Red Crescent Societies in different countries can generate awareness and inform all actors, including individual donors, about acceptable and appropriate donations for survivors.

Depending on whether disasters occur in developed or developing countries, domestic donors must be informed and encouraged to donate appropriate relief items. For example, the actions of state and local governments, as well as volunteer organizations responsible for relief operations after the 2013 Colorado flood, contributed to reducing material overload and convergence. The actions included implementing an electronic messaging method to encourage financial or high-priority material donations and using effective donations management strategies to discourage people from donating unnecessary items. In addition, the software to manage coordinate material donations provided the ability to match offers and needs in this particular situation (Arnette and Zobel 2016). Of course, this may not be applicable for developing countries because of lack of appropriate technology and resources. However, public education and awareness campaigns may be helpful for people of these countries. As indicated in Chapter 5, cash distributed among the survivors of the 2005 Pakistan earthquake helped reduce the flow of unnecessary relief items in earthquake-affected areas of the country.

Like the materials convergence, convergence of volunteers from foreign countries also needs to be controlled. Although trained and skilled volunteers are critical in the aftermath of a disaster, too many of them in a particular disaster site can create several logistical problems that interfere with timely distribution of aid. More specifically, participation of untrained and unskilled overseas volunteer can become a liability after a disaster. Thus, not all of them should be allowed into the disaster-affected countries. Their flow must be reduced at both the exit and entry points, and only those volunteers who have adequate training, experience, and who are affiliated with reputed relief organizations should be allowed to participate in disaster relief operations.

Alternatively, a quota system for foreign volunteers could be introduced and their number should be determined by the appropriate authorities given the severity of the disaster and available resources. International relief agencies like the IFRC should encourage volunteers to help disaster victims in their own country instead of traveling to foreign countries. Convergence of too much information has not yet created any serious problem in conducting relief operations; however, care must be exercised so that it does not happen.

Lack of coordination and collaboration among the parties and agencies involved in disaster relief operations still remains a serious problem and has been for many decades (Kent 1987; Rey 2001). Although many attempts have been made to improve coordination at various scales and levels, almost all known disaster relief operations suffer from this problem. For example, the United Nations has established various committees and offices (e.g., OCHA, IASC, Central Emergency Response Fund (CERF), and CAP) to improve coordination within the relief community. However, among the principal reasons for lack of cooperation is the emergence of many relief organizations, participation of a large number of such organizations in a particular relief operation, intense competition for funding among these organizations, relief organizations' reluctance to share information with other organizations, increasing scarcity of global resources, lack of accountability and high cost of coordination, lack of trust, staff turnover, unilateral donor actions and communication or language challenges between local and foreign organizations.[3] These can be considered barriers to coordination.

Advantages of coordination are discussed in Chapter 2 highlighting that with collaboration of other relevant agencies, the International Federation of Red Cross and Red Crescent Societies developed a module to improve coordination among organizations participating in disaster relief operations (IFRC 2000). The module acknowledges that no single organization can work alone to address magnitude and the complexity of the need associated with disaster response. Accordingly, the IFRC (2000) suggests 11 actions that facilitate coordination. These actions are listed in Table 6.1. Although these suggestions were outlined almost two decades ago, they are still valid. Also, all concerned organizations need to build both formal and informal mechanisms as well as maintain conversation, for improving and increasing communication among themselves.

Improvement in many aspects of coordination and communication among authorities and agencies is required to improve disaster relief operations. Several authors (e.g., Ito et al. 2014; Mulyono and Ishida 2014) suggest improved performance of relief operations is achieved with better management of logistics and inventory activities. To quickly deliver essential items like foods, drinking water, clothes, and medicines to disaster survivors and evacuees requires pre-positioning of necessary

Table 6.1 Actions that facilitate coordination

1. Identify objectives and mutual interests
2. Need effective negotiation skills
3. Seek and secure funds for coordination
4. Discuss positive aspects of coordination and improve awareness of interdependence
5. Standardize organizational activities
6. Strategically select the number and participatory organizations
7. Build personal relationships through trust
8. Invite effective leaders and others to join the coordination body
9. Promote dynamic and effective leadership
10. Ensure local decision-making capacity
11. Improve skills, knowledge, and experience associated with coordination

Source Prepared by the author based on IFRC (2000, 16–17)

relief items at designated shelters or warehouses. This requires an inventory of available items in these facilities which must take place before the disaster. Inventorying and pre-positioning materials will address victims' needs in a timely and efficient manner.

Future Research

Studies are needed to measure the impact of a humanitarian response that is focused on providing relief benefit efficiently, which is critical to improve future disaster responses and efficacy (Kirsch et al. 2012). More research is also needed about the organizational cultures and constraints of relief agencies themselves. This should move beyond criticizing to offer practical solutions as to how to address the gap between research and practice, perhaps focusing on how bureaucratic barriers might be transcended to encourage situational flexibility and generate genuine grass-roots participation. This should include the moral, social, political, and economic values that the recipients of aid attach to the items being provided in relief (Prendergast 1996).

Despite its importance in effectively communicating and disseminating crucial information related to disaster response and operations in real time, social media is still a relatively under-researched area in hazard studies. Future research must fill this gap by addressing how this modern technology mobilizes agents for disaster response and operations, how the technology makes both efforts more effective, what the problems are of using this technology, and many other aspects associated with its use.

For major disasters, Flash Appeals are generally made by more than one entity, which raises two important questions: Should this be allowed to create confusion among providers of disaster aid? How effective are these Flash Appeals in convincing providers to donate to survivors of disasters and complex emergencies? Hardly, any empirical studies are available to provide answers to these questions. Future studies could fill this research gap.

Most of the examples of disaster relief operation cited in this book are drawn from Asia, and North and Middle America, including the

Caribbean region. This does not necessarily mean that other countries of other continents/regions are not prone to natural disasters and complex emergencies but rather that relevant information on disaster relief efforts is either non-existent or very limited. This is a serious research gap that hazard and disaster researchers need to fill.

In this context, disaster relief experience of small island countries and territories, for example, must be noted. The United Nations grouped them as Small Island Developing States (SIDS). These states are located in the Atlantic, Indian, and Pacific Oceans as well as in the Caribbean, Mediterranean, and China Seas. Many of the SIDS experience natural disasters like hurricanes and tsunamis. For example, in 2017, several SIDS in the Caribbean were devastated by several hurricanes. Unfortunately, virtually no information regarding relief efforts undertaken in these countries and territories exists. Because of their small size with respect to both area and population, limited financial resources, weak infrastructure, and a strong reliance on natural resources, data about emergency relief operations in these countries would provide valuable and new insights regarding the problems associated with arranging, collecting, distributing, and managing of emergency aid to affected populations. Clearly, future research on disaster relief in SIDS would be valuable.

Despite a lack of comprehensive research studies on the designated areas, this book nonetheless offers detailed insights into the dynamics of disaster relief to a wide range of readers: academics and students from a variety of disciplines (e.g., geographers, sociologists, anthropologists, psychologists, economists, and engineers) as well as disaster managers and planners, policy-makers, and a multiplicity of organizations and agencies involved in international disaster relief interventions. It provides a multiscale view of emergency relief efforts over the past several decades and as such it updates the current disaster response and relief literature.

This book identifies the purposes of and channels for disaster relief and provides a historical overview of international disaster aid in relation to the field of natural hazards and disasters. It also identifies the intentions and potential pitfalls of aid distribution in a discussion of the actors and processes involved. Furthermore, the book offers a set of success/failure indicators based on literature review as well as empirical research conducted in several countries on different natural disasters. The research culminates in lessons learned and offers recommendations for a path toward more useful relief operations and ultimately concludes that the comprehensive discussion of the evolution of actors, intentions, processes, and challenges is applicable to any type of natural disaster or complex emergency.

Notes

1. Some of these organizations/agencies were briefly introduced in previous chapters, while the descriptions of others are provided below.
2. This is closely associated with Putnam's discussion of building local-level capacity and strengthening local social capital (Putnam 1993). This is also aligned with the "target group" approach (Ahmed et al. 1997).
3. Each of the disaster relief operations covered in Chapter 4 was participated in by more than 150 foreign and domestic organizations and agencies.

References

Ahmed, M., A. Capistrano, and M. Hossain. 1997. Experience of Partnership Models for the Co-management of Bangladesh Fisheries. *Fish Management Ecology* 4: 233–248.

Arnette, A.N., and C.W. Zobel. 2016. Investigation of Material Convergence in the September 2013 Colorado Floods. *Natural Hazards Review* 17 (2): 05016001.

Balcik, B., B.M. Beamon, C. Krejci, and M. Ramirez. 2010. Coordination in Humanitarian Relief Chains: Practices, Challenges and Opportunities. *International Journal of Production Economics* 126 (1): 22–34.

Flint, M., and H. Goyder. 2006. *Funding the Tsunami Response*. London: Tsunami Evaluation Coalition.

Hebblethwaite, B. 2011. Haiti Earthquake (2010). In *Encyclopedia of Disaster Relief*, ed. K.B. Penuel and M. Statler, 258–262. Newbury Park: Sage.

Henry, D. 2005. Anthropological Contributions to the Study of Disasters. In *Disciplines, Disasters and Emergency Management: The Convergence and Divergence of Concepts, Issues and Trends from the Research Literature*, ed. D. McEntire and W. Blanchard. Emittsburg, MD: Federal Emergency

Management Agency. http://training.fema.gov/emiweb/edu/ddemtextbook.asp. Last Accessed 23 Nov 2017.

IFRC (International Federation of Red Cross and Red Crescent Societies). 2000. *Improving Coordination*. Geneva: IFRC.

IFRC (International Federation of Red Cross and Red Crescent Societies). 2013. *World Disasters Report: Focus on Technology and the Future of Humanitarian Action*. Geneva: IFRC.

Islam, R., and G. Walkerden. 2015. How do Links Between Households and NGOs Promote Disaster Resilience and Recovery? A Case Study of Linking Social Networks on the Bangladesh Coast. *Natural Hazards* 78 (3): 1707–1727.

Ito, H., W. Wisetjindawat, and M. Yokomatsu. 2014. Improving the Operational Efficiency of Humanitarian Logistics in the Aftermath of a Large-Scale Disaster. *Journal of Integrated Disaster Risk Management* 4 (2): 142–155.

Kent, R.C. 1987. *Anatomy of Disaster Relief: The International Network in Action*. London: Pinter Publishers.

Kirsch, T.D., P. Perrin, F.M. Burkle, W. Canny, S. Purdin, W. Len, and L. Sauer. 2012. Requirements for Independent Community-Based Quality Assessment and Accountability Practices in Humanitarian Assistance and Disaster Relief Activities. *Prehospital and Disaster Medicine* 27 (3): 280–285.

Mulyono, N.B., and Y. Ishida. 2014. Humanitarian Logistics and Inventory Model Based on Probabilistic Cellular Automata. *International Journal of Innovative Computing, Information, and Control* 10 (1): 357–372.

Paul, B.K., B. Acharya, and K. Ghimire. 2017. Effectiveness of Earthquakes Relief Efforts in Nepal: Opinions of the Survivors. *Natural Hazard* 85 (2017): 1169–1188.

Prendergast, J. 1996. *Frontline Diplomacy: Humanitarian Aid and Conflict in Africa*. Boulder, CO: Lynne Rienner.

Putnam, R.D. 1993. *Making Democracy Work: Civic Traditions in Modern Italy*. Princeton: Princeton University Press.

Rey, F. 2001. The Complex Nature of Actors in Humanitarian Action and the Challenge of Coordination. *In Reflections on Humanitarian Action: Principles, Ethics and Contradictions*, ed. Humanitarian Studies Unit. London: TNI/Pluto Press.

Taylor, G., A. Stoddard, A. Harmer, K. Haver, and P. Harvey. 2012. *The State of the Humanitarian System* (2012 Edition). London: Overseas Development Institute.

USAID (U.S. Agency for International Development). 2002. Mozambique Flood Resettlement Grant Activity. Washington, DC: USAID.

Index

B. K. Paul, *Disaster Relief Aid*,
https://doi.org/10.1007/978-3-319-77282-0

I

The manufacturer's authorised representative in the EU is Springer Nature Customer Service Centre GmbH, Europaplatz 3, 69115 Heidelberg, Germany. If you have any concerns regarding our products, please contact ProductSafety@springernature.com

Printed and bound by CPI Group (UK) Ltd, Croydon, CR0 4YY
30/09/2026
02225660-0010